Special Publication 500-291

NIST Cloud Computing
Standards Roadmap – Version 1.0

NIST Cloud Computing Standards Roadmap
Working Group
Michael Hogan
Fang Liu
Annie Sokol
Jin Tong

INTERNETWORK TECHNOLOGIES

Computer Security Division
Information Technology Laboratory
National Institute of Standards and Technology
Gaithersburg, MD 20899-8930

July 2011

U. S. Department of Commerce

Gary Locke, Secretary

National Institute of Standards and Technology

Patrick D. Gallagher, Director

Reports on Computer Systems Technology

The Information Technology Laboratory (ITL) at the National Institute of Standards and Technology (NIST) promotes the U.S. economy and public welfare by providing technical leadership for the nation's measurement and standards infrastructure. ITL develops tests, test methods, reference data, proof of concept implementations, and technical analyses to advance the development and productive use of information technology. ITL's responsibilities include the development of technical, physical, administrative, and management standards and guidelines for the cost-effective security and privacy of sensitive unclassified information in federal computer systems. This Special Publication 500-series reports on ITL's research, guidance, and outreach efforts in Information Technology and its collaborative activities with industry, government, and academic organizations.

National Institute of Standards and Technology Special Publication 500-291

Natl. Inst. Stand. Technol. Spec. Publ. 500-291, 83 pages (5 July 2011)

DISCLAIMER

This document has been prepared by the National Institute of Standards and Technology (NIST) and describes standards research in support of the NIST Cloud Computing Program.

Certain commercial entities, equipment, or material may be identified in this document in order to describe a concept adequately. Such identification is not intended to imply recommendation or endorsement by the National Institute of Standards and Technology, nor is it intended to imply that these entities, materials, or equipment are necessarily the best available for the purpose.

Acknowledgements

This document emerged from a protracted period of discussions with a number of stakeholders including the NIST Cloud Computing Standards Roadmap working group, chaired by Michael Hogan and Annie Sokol of the U.S. Department of Commerce, National Institute of Standards and Technology. The document contains input from members of the NIST Cloud Computing Working Groups: Reference Architecture Working Group led by Robert Bohn and John Messina; Standards to Acceleration to Jumpstart the Adoption of Cloud Computing (SAJACC) led by Lee Badger; Security Working Group led by Lee Badger; and Business Use Cases Working Group led by Fred Whiteside.

The staff at Knowcean Consulting including John Mao, Jin Tong, and Fang Liu, provided commendable contribution and guidance.

Table of Contents

EXECUTIVE SUMMARY 10

1 INTRODUCTION 12
   1.1 BACKGROUND 12
   1.2 NIST CLOUD COMPUTING VISION 12
   1.3 NIST CLOUD COMPUTING STANDARDS ROADMAP WORKING GROUP 13
   1.4 HOW THIS REPORT WAS PRODUCED 13

2 THE NIST DEFINITION OF CLOUD COMPUTING 14

3 CLOUD COMPUTING REFERENCE ARCHITECTURE 16
   3.1 OVERVIEW 16
   3.2 CLOUD CONSUMER 17
   3.3 CLOUD PROVIDER 19
   3.4 CLOUD AUDITOR 24
   3.5 CLOUD BROKER 25
   3.6 CLOUD CARRIER 25

4 CLOUD COMPUTING USE CASES 26
   4.1 BUSINESS USE CASES 26
   4.2 TECHNICAL USE CASES 26
   4.3 DEPLOYMENT SCENARIO PERSPECTIVE 27

5 CLOUD COMPUTING STANDARDS 31
   5.1 INFORMATION AND COMMUNICATION TECHNOLOGIES (IT) STANDARDS LIFE CYCLE 31
   5.2 CATEGORIZING THE STATUS OF STANDARDS 32
   5.3 CLOUD COMPUTING STANDARDS FOR INTEROPERABILITY 33
   5.4 CLOUD COMPUTING STANDARDS FOR PORTABILITY 36
   5.5 CLOUD COMPUTING STANDARDS FOR SECURITY 39

6 CLOUD COMPUTING STANDARDS MAPPING AND GAP ANALYSIS 42
   6.1 SECURITY STANDARDS MAPPING 42
   6.2 INTEROPERABILITY STANDARDS MAPPING 44
   6.3 PORTABILITY STANDARDS MAPPING 45
   6.4 USE CASE ANALYSIS 45

7 CLOUD COMPUTING STANDARDS GAPS AND USG PRIORITIES 50
   7.1 AREAS OF STANDARDIZATION GAPS 51
   7.2 STANDARDIZATION PRIORITIES BASED ON USG PRIORITIES TO STANDARDIZATION PRIORITIES BASED ON USG CLOUD COMPUTING ADOPTION PRIORITIES 53

8 CONCLUSIONS AND RECOMMENDATIONS 56

BIBLIOGRAPHY 58

APPENDIX A – NIST SPECIAL PUBLICATIONS RELEVANT TO CLOUD COMPUTING 59

APPENDIX B – DEFINITIONS 60

APPENDIX C – ACRONYMS 63
APPENDIX D – STANDARDS DEVELOPING ORGANIZATIONS 66
APPENDIX E – CONCEPTUAL MODELS AND ARCHITECTURES 72
APPENDIX F – EXAMPLES OF USG CRITERIA FOR SELECTION OF STANDARDS 73

List of Figures and Tables

Figure 1 – Interactions between the Actors in Cloud Computing 17

Figure 2 – Example of Services Available to a Cloud Consumer 18

Figure 3 – Cloud Provider: Major Activities 20

Figure 4 – Cloud Provider: Service Orchestration 21

Figure 5 – Cloud Provider: Cloud Service Management 23

Figure 6 – High-Level Generic Scenarios 27

Figure 7 – IT Standards Life Cycle 31

Figure 8 – Cloud Service Presents an Interface to Each Category 33

Figure 9 – IaaS Interface 34

Figure 10 – PaaS Interface 34

Figure 11 – SaaS Interface 35

Figure 12 – The Combined Conceptual Reference Diagram 42

Figure 13 – DoD DISR Standards Selection Process 76

Table 1 – Actors in Cloud Computing 17

Table 2 – Cloud Consumer and Cloud Provider 18

Table 3 – Deployment Cases for High-Level Scenarios 28

Table 4 – Standards Maturity Model 32

Table 5 – Security: Categorization 44

Table 6 – Interoperability: Categorization 45

Table 7 – Portability: Categorization 45

Table 8 – Areas of Standardization Gaps and Standardization Priorities 51

Table 9 – DoD Selection Criteria and Description Summary 75

Table 10 – DoD Standards Sources Preferences 76

NIST Cloud Computing Standards Roadmap

Executive Summary

The National Institute of Standards and Technology (NIST) has been designated by Federal Chief Information Officer (CIO) Vivek Kundra to accelerate the federal government's secure adoption of cloud computing by leading efforts to develop standards and guidelines in close consultation and collaboration with standards bodies, the private sector, and other stakeholders.

The NIST Cloud Computing Standards Roadmap Working Group has surveyed the existing standards landscape for security, portability, and interoperability standards/models/studies/use cases, etc., relevant to cloud computing. Using this available information, current standards, standards gaps, and standardization priorities are identified in this document.

The NIST Definition of Cloud Computing identified cloud computing as a model for enabling ubiquitous, convenient, on-demand network access to a shared pool of configurable computing resources (e.g., networks, servers, storage, applications, and services) that can be rapidly provisioned and released with minimal management effort or service provider interaction.

As an extension to the above NIST cloud computing definition, a NIST cloud computing reference architecture has been developed by the NIST Cloud Computing Reference Architecture and Taxonomy Working Group that depicts a generic high-level conceptual model for discussing the requirements, structures and operations of cloud computing. It contains a set of views and descriptions that are the basis for discussing the characteristics, uses, and standards for cloud computing, and relates to a companion cloud computing taxonomy.

Cloud computing use cases describe the consumer requirements in using cloud computing service offerings. Using existing use cases, this document analyzes how existing cloud-related standards fit the needs of federal cloud consumers and identifies standardization gaps.

Cloud computing standards are already available in support of many of the functions and requirements for cloud computing. While many of these standards were developed in support of pre-cloud computing technologies, such as those designed for web services and the Internet, they also support the functions and requirements of cloud computing. Other standards are now being developed in specific support of cloud computing functions and requirements, such as virtualization.

To assess the state of standardization in support of cloud computing, the NIST Cloud Computing Standards Roadmap Working Group has compiled an Inventory of Standards Relevant to Cloud Computing. This inventory is being maintained and will be used to update this document as necessary. Using the taxonomy developed by the NIST Cloud Computing Reference Architecture and Taxonomy Working Group, cloud computing relevant standards have been mapped to the requirements of portability, interoperability, and security.

Present areas with standardization gaps include: SaaS functional interfaces; SaaS self-service management interfaces; PaaS functional interfaces; business support / provisioning / configuration; and security and privacy. Present standardization areas of priority to the federal government include: security auditing and compliance; identity and access management; SaaS application specific data and metadata; and resource description and discovery.

There is a fast-changing landscape of cloud computing-relevant standardization under way in a number of Standards Developing Organizations (SDOs). While there are only a few approved cloud computing specific standards at present, federal agencies should be encouraged to participate in specific cloud computing standards development projects that support their priorities in cloud computing services. Specific recommendations are:

Recommendation 1 – Contribute Agency Requirements
Agencies should contribute clear and comprehensive user requirements for cloud computing standards projects.

Recommendation 2 – Participate in Standards Development
Agencies should actively participate in cloud computing standards development projects that are of high priority to their agency missions.

Recommendation 3 – Encourage Testing to Accelerate Technically Sound Standards-Based Deployments
Agencies should support the concurrent development of conformity and interoperability assessment schemes to accelerate the development and use of technically sound cloud computing standards and standards-based products, processes and services.

Recommendation 4 – Specify Cloud Computing Standards
Agencies should specify cloud computing standards in their procurements and grant guidance when multiple vendors offer standards-based implementations and there is evidence of successful interoperability testing. In such cases, agencies should ask vendors to show compliance to the specified standards.

Recommendation 5 – United States Government (USG) – Wide Use of Cloud Computing Standards
To support USG requirements for interoperability, portability, and security in cloud computing, the Federal Standards and Technology Working Group chaired by NIST and complimentary to the Fed CIO Council Cloud Computing Executive Steering Committee (CCESC) and Cloud First Task Force should recommend specific cloud computing standards and best practices for USG-wide use.

Recommendation 6 – Dissemination of Information on Cloud Computing Standards
A listing of standards relevant to cloud computing should be posted and maintained by NIST.

1 Introduction

1.1 Background

U.S. laws and associated policy require federal agencies to use international, voluntary consensus standards in their procurement and regulatory activities, except where inconsistent with law or otherwise impractical.[1]

NIST has been designated by Federal CIO Vivek Kundra to accelerate the federal government's secure adoption of cloud computing by leading efforts to develop standards and guidelines in close consultation and collaboration with standards bodies, the private sector, and other stakeholders.

The NIST Cloud Computing Program was formally launched in November 2010 and was created to support the federal government effort to incorporate cloud computing as a replacement for, or enhancement to, traditional information system and application models where appropriate. The NIST Cloud Computing Program operates in coordination with other federal cloud computing implementation efforts (CIO Council/Information Security and Identity Management Committee [ISIMC], etc.) and is integrated with Federal CIO Vivek Kundra's 25-point IT Implementation Plan for the federal government. NIST has created the following working groups in order to provide a technically oriented strategy and standards-based guidance for the federal cloud computing implementation effort:

   Cloud Computing Reference Architecture and Taxonomy Working Group
   Cloud Computing Standards Acceleration to Jumpstart Adoption of Cloud Computing (SAJACC) Working Group
   Cloud Computing Security Working Group
   Cloud Computing Standards Roadmap Working Group
   Cloud Computing Target Business Use Cases Working Group

1.2 NIST Cloud Computing Vision

NIST's long-term goal is to provide leadership and guidance around the cloud computing paradigm to catalyze its use within industry and government. NIST aims to shorten the adoption cycle, which will enable near-term cost savings and increased ability to quickly create and deploy safe and secure enterprise solutions. NIST aims to foster cloud computing practices that support interoperability, portability, and security requirements that are appropriate and achievable for important usage scenarios.

The NIST area of focus is technology, and specifically, interoperability, portability, and security requirements,standards, and guidance. The intent is to use the standards strategy to

---

[1] Trade Agreements Act of 1979, as amended (TAA) , the National Technology Transfer and Advancement Act (NTTAA), and The Office of Management and Budget (OMB) Circular A-119 Revised: Federal Participation in the Development and Use of Voluntary Consensus Standards and in Conformity Assessment Activities

prioritize NIST tactical projects which support USG agencies in the secure and effective adoption of the cloud computing model to support their missions. The expectation is that the set of priorities will be useful more broadly by industry, SDOs, cloud adopters, and policy makers.

1.3 NIST Cloud Computing Standards Roadmap Working Group

Standards Developing Organizations and others have and are developing supporting cloud computing documents to include standards, conceptual models, reference architectures, and standards roadmaps to facilitate communication, data exchange, and security for cloud computing and its application. Still other standards are emerging to focus on technologies that support cloud computing, such as virtualization. The NIST Cloud Computing Standards Roadmap Working Group is leveraging this existing, publicly available work, plus the work of the other NIST working groups, to develop a NIST Cloud Computing Standards Roadmap that can be incorporated into the NIST USG Cloud Computing Technology Roadmap.

1.4 How This Report Was Produced

The NIST Cloud Computing Standards Roadmap Working Group (CCSRWG) has surveyed the existing standards landscape for security, portability, and interoperability standards / models / studies / use cases, etc., relevant to cloud computing. Using this available information, standards, standards gaps or overlaps, and standardization priorities have been identified.

Future editions of this report may consider additional areas, such as maintainability, usability, reliability, and resiliency.

2 The NIST Definition of Cloud Computing[2]

Cloud computing is a model for enabling ubiquitous, convenient, on-demand network access to a shared pool of configurable computing resources (e.g., networks, servers, storage, applications, and services) that can be rapidly provisioned and released with minimal management effort or service provider interaction. This cloud model promotes availability and is composed of five essential characteristics, three service models, and four deployment models.

Essential Characteristics:

On-demand self-service. A consumer can unilaterally provision computing capabilities, such as server time and network storage, as needed automatically without requiring human interaction with each service's provider.

Broad network access. Capabilities are available over the network and accessed through standard mechanisms that promote use by heterogeneous thin or thick client platforms (e.g., mobile phones, laptops, and personal digital assistants [PDAs]).

Resource pooling. The provider's computing resources are pooled to serve multiple consumers using a multi-tenant model, with different physical and virtual resources dynamically assigned and reassigned according to consumer demand. There is a sense of location independence in that the customer generally has no control or knowledge over the exact location of the provided resources but may be able to specify location at a higher level of abstraction (e.g., country, state, or datacenter). Examples of resources include storage, processing, memory, network bandwidth, and virtual machines.

Rapid elasticity. Capabilities can be rapidly and elastically provisioned, in some cases automatically, to quickly scale out and rapidly released to quickly scale in. To the consumer, the capabilities available for provisioning often appear to be unlimited and can be purchased in any quantity at any time.

Measured Service. Cloud systems automatically control and optimize resource use by leveraging a metering capability[3] at some level of abstraction appropriate to the type of service (e.g., storage, processing, bandwidth, and active user accounts). Resource usage can be monitored, controlled, and reported, providing transparency for both the provider and consumer of the utilized service.

---

[2] NIST Special Publication 800-145, *The NIST Definition of Cloud Computing*, http://csrc.nist.gov/publications/drafts/800-145/Draft-SP-800-145_cloud-definition.pdf
[3] Typically through a pay-per-use business model.

Service Models:

Cloud Software as a Service (SaaS). The capability provided to the consumer is to use the provider's applications running on a cloud infrastructure. The applications are accessible from various client devices through a thin client interface such as a Web browser (e.g., Web-based email). The consumer does not manage or control the underlying cloud infrastructure including network, servers, operating systems, storage, or even individual application capabilities, with the possible exception of limited user-specific application configuration settings.

Cloud Platform as a Service (PaaS). The capability provided to the consumer is to deploy onto the cloud infrastructure consumer-created or acquired applications created using programming languages and tools supported by the provider. The consumer does not manage or control the underlying cloud infrastructure including network, servers, operating systems, or storage, but has control over the deployed applications and possibly application hosting environment configurations.

Cloud Infrastructure as a Service (IaaS). The capability provided to the consumer is to provision processing, storage, networks, and other fundamental computing resources where the consumer is able to deploy and run arbitrary software, which can include operating systems and applications. The consumer does not manage or control the underlying cloud infrastructure but has control over operating systems, storage, deployed applications, and possibly limited control of select networking components (e.g., host firewalls).

Deployment Models:

Private cloud. The cloud infrastructure is operated solely for an organization. It may be managed by the organization or a third party and may exist on premise or off premise.

Community cloud. The cloud infrastructure is shared by several organizations and supports a specific community that has shared concerns (e.g., mission, security requirements, policy, and compliance considerations). It may be managed by the organizations or a third party and may exist on premise or off premise.

Public cloud. The cloud infrastructure is made available to the general public or a large industry group and is owned by an organization selling cloud services.

Hybrid cloud. The cloud infrastructure is a composition of two or more clouds (private, community, or public) that remain unique entities but are bound together by standardized or proprietary technology that enables data and application portability (e.g., cloud bursting for load balancing between clouds).

3 Cloud Computing Reference Architecture

The NIST cloud computing definition is widely accepted and valuable in providing a clear understanding of cloud computing technologies and cloud services. The NIST cloud computing reference architecture presented in this clause is a natural extension to the NIST cloud computing definition.

The NIST cloud computing reference architecture is a generic high-level conceptual model that is a powerful tool for discussing the requirements, structures, and operations of cloud computing. The model is not tied to any specific vendor products, services, or reference implementation, nor does it define prescriptive solutions that inhibit innovation. It defines a set of actors, activities, and functions that can be used in the process of developing cloud computing architectures, and relates to a companion cloud computing taxonomy. It contains a set of views and descriptions that are the basis for discussing the characteristics, uses and standards for cloud computing.

The NIST cloud computing reference architecture focuses on the requirements of what cloud service provides, *not* on a design that defines a solution and its implementation. It is intended to facilitate the understanding of the operational intricacies in cloud computing. The reference architecture does not represent the system architecture of a specific cloud computing system; instead, it is a tool for describing, discussing, and developing the system-specific architecture using a common framework of reference.

The design of the NIST cloud computing reference architecture serves the objectives to: illustrate and understand various cloud services in the context of an overall cloud computing conceptual model; provide technical references to USG agencies and other consumers to understand, discuss, categorize, and compare cloud services; and communicate and analyze security, interoperability, and portability candidate standards and reference implementations.

3.1 Overview

The NIST cloud computing reference architecture defines five major *actors*: *cloud consumer, cloud provider, cloud auditor, cloud broker,* and *cloud carrier*. Each actor is an entity (a person or an organization) that participates in a transaction or process and/or performs tasks in cloud computing. Table 1 briefly lists the five major actors defined in the NIST cloud computing reference architecture.

Figure 1 shows the interactions among the actors in the NIST cloud computing reference architecture. A cloud consumer may request cloud services from a cloud provider directly or via a cloud broker. A cloud auditor conducts independent audits and may contact the others to collect necessary information. The details will be discussed in the following sections and be presented as successive diagrams in increasing levels of detail.

Actor	Definition
Cloud Consumer	Person or organization that maintains a business relationship with, and uses service from, *Cloud Providers*.
Cloud Provider	Person, organization, or entity responsible for making a service available to *Cloud Consumers*.
Cloud Auditor	A party that can conduct independent assessment of cloud services, information system operations, performance, and security of the cloud implementation.
Cloud Broker	An entity that manages the use, performance, and delivery of cloud services, and negotiates relationships between *Cloud Providers* and *Cloud Consumers*.
Cloud Carrier	The intermediary that provides connectivity and transport of cloud services from *Cloud Providers* to *Cloud Consumers*.

Table 1 – Actors in Cloud Computing

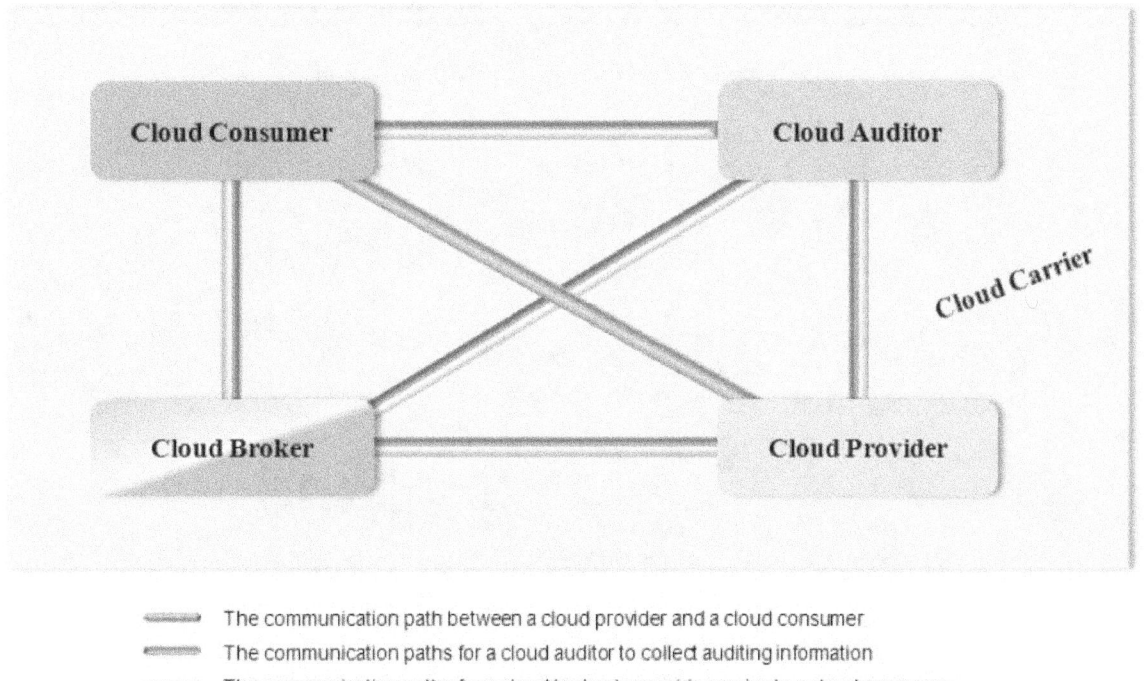

— The communication path between a cloud provider and a cloud consumer
— The communication paths for a cloud auditor to collect auditing information
— The communication paths for a cloud broker to provide service to a cloud consumer

Figure 1 – Interactions between the Actors in Cloud Computing

3.2 Cloud Consumer

The cloud consumer is the ultimate stakeholder that the cloud computing service is created to support. A cloud consumer represents a person or organization that maintains a business relationship with, and uses the service from, a cloud provider. A cloud consumer browses the service catalog from a cloud provider, requests the appropriate service, sets up service contracts with the cloud provider, and uses the service. The cloud consumer may be billed for the service provisioned, and needs to arrange payments accordingly. Depending on the services requested, the activities and usage scenarios can be different among cloud consumers, as shown in Table 2. Some example usage scenarios are listed in Figure 2.

Type	Consumer Activities	Provider Activities
SaaS	Uses application/service for business process operations.	Installs, manages, maintains, and supports the software application on a cloud infrastructure.
PaaS	Develops, tests, deploys, and manages applications hosted in a cloud environment.	Provisions and manages cloud infrastructure and middleware for the platform consumers; provides development, deployment, and administration tools to platform consumers.
IaaS	Creates/installs, manages, and monitors services for IT infrastructure operations.	Provisions and manages the physical processing, storage, networking, and the hosting environment and cloud infrastructure for IaaS consumers.

Table 2 – Cloud Consumer and Cloud Provider

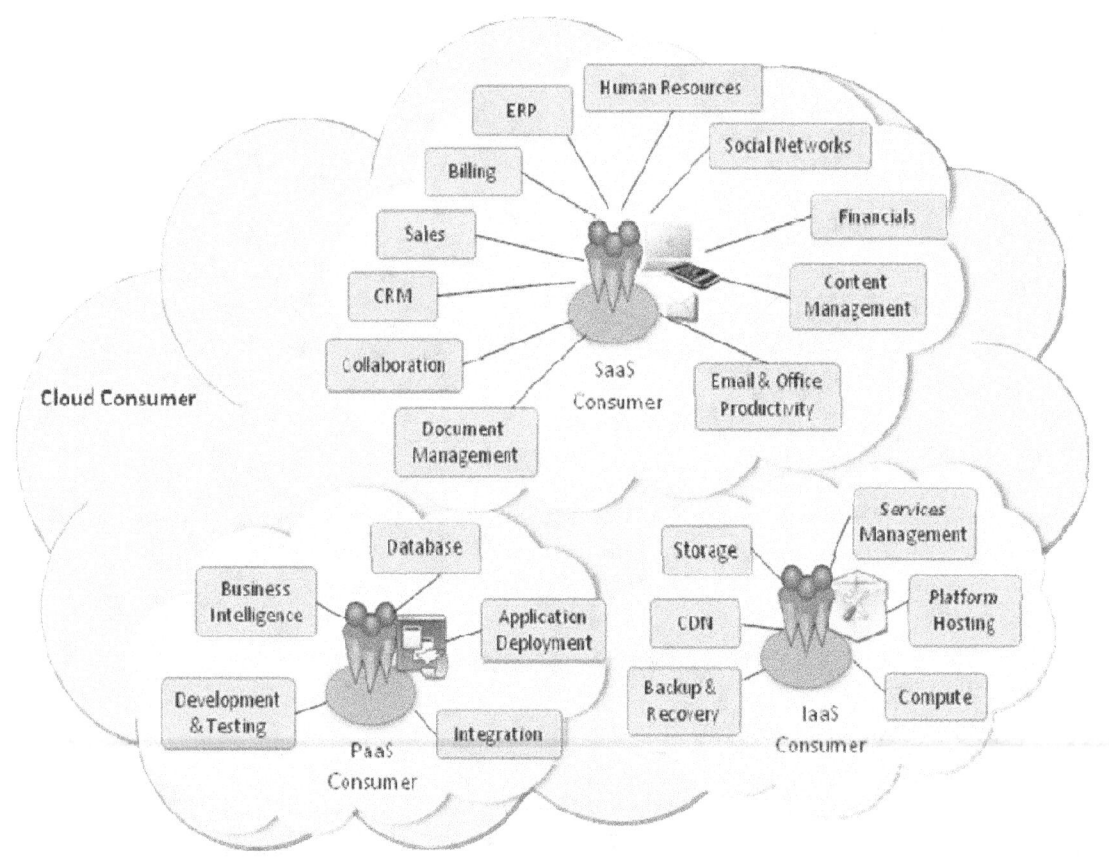

Figure 2 – Example of Services Available to a Cloud Consumer

SaaS applications are usually deployed as hosted services and are accessed via a network connecting SaaS consumers and providers. The consumers of SaaS can be organizations that provide their members with access to software applications, end users who directly use software applications, or software application administrators who configure applications for end users. SaaS consumers access and use applications on demand, and can be billed on the number of consumers or the amount of consumed services. The latter can be measured in terms of the time in use, the network bandwidth consumed, or the amount/duration of data stored.

Cloud consumers who use PaaS can employ the tools and execution resources provided by cloud providers for the purpose of developing, testing, deploying, and managing applications hosted in a cloud environment. PaaS consumers can be application developers who design and implement application software, application testers who run and test applications in various cloud-based environments, application deployers who publish applications into the cloud, and application administrators who configure and monitor application performance on a platform. PaaS consumers can be billed by the number of consumers, the type of resources consumed by the platform, or the duration of platform usage.

IaaS clouds provision consumers the capabilities to access virtual computers, network-accessible storage, network infrastructure components, and other fundamental computing resources, on which consumers can deploy and run arbitrary software. The consumers of IaaS can be system developers, system administrators, and information technology (IT) managers who are interested in creating, installing, managing and monitoring services for IT infrastructure operations. IaaS consumers are provisioned with the capabilities to access these computing resources, and are billed for the amount of resources consumed.

3.3 Cloud Provider

A cloud provider can be a person, an organization, or an entity responsible for making a service available to cloud consumers. A cloud provider builds the requested software/platform/infrastructure services, manages the technical infrastructure required for providing the services, provisions the services at agreed-upon service levels, and protects the security and privacy of the services. As illustrated in Table 2, cloud providers undertake different tasks for the provisioning of the various service models.

For Cloud Software as a Service, the cloud provider deploys, configures, maintains, and updates the operation of the software applications on a cloud infrastructure so that the services are provisioned at the expected service levels to cloud consumers. The provider of SaaS assumes most of the responsibilities in managing and controlling the applications and the infrastructure, while the cloud consumers have limited administrative control of the applications.

For Cloud Platform as a Service, the cloud provider manages the cloud infrastructure for the platform, and provisions tools and execution resources for the platform consumers to develop, test, deploy, and administer applications. Consumers have control over the applications and possibly the hosting environment settings, but cannot access the infrastructure underlying the platform including network, servers, operating systems, or storage.

For Cloud Infrastructure as a Service, the cloud provider provisions the physical processing, storage, networking, and other fundamental computing resources, as well as manages the hosting environment and cloud infrastructure for IaaS consumers. Cloud consumers deploy and run applications, have more control over the hosting environment and operating systems, but do not manage or control the underlying cloud infrastructure (e.g., the physical servers, network, storage, hypervisors, etc.).

The activities of cloud providers can be discussed in greater detail from the perspectives of *Service Deployment, Service Orchestration, Cloud Service Management, Security* and *Privacy*.

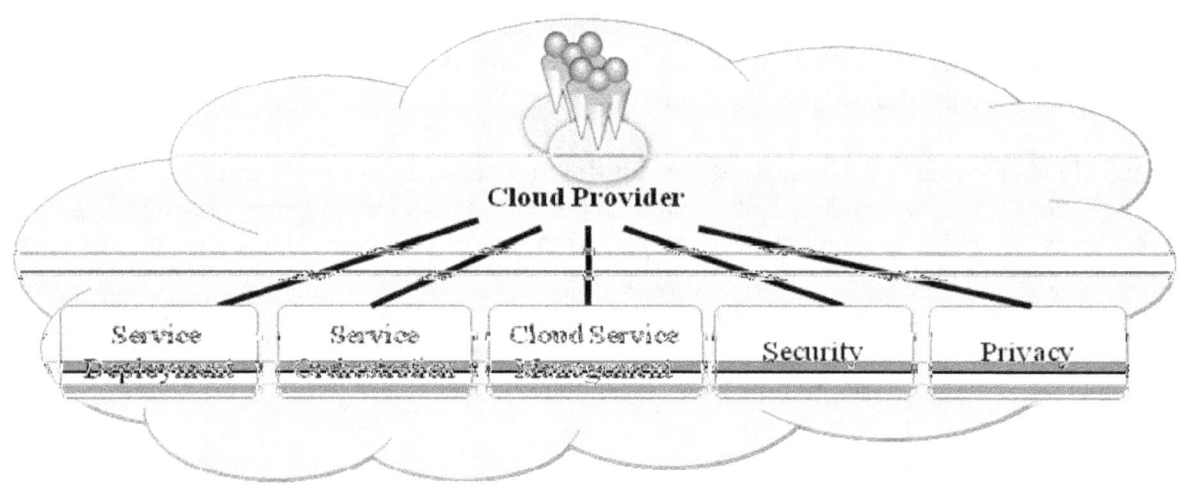

Figure 3 – Cloud Provider: Major Activities

3.3.1 Service Deployment

As identified in the NIST cloud computing definition, a cloud infrastructure may be operated in one of the following deployment models: *public cloud*, *private cloud*, *community cloud*, or *hybrid cloud*. For the details related to the controls and management in the cloud, we refer the readers to the NIST Special Publication 800-146, *NIST Cloud Computing Synopsis and Recommendations*.

A public cloud is one in which the cloud infrastructure and computing resources are made available to the general public over a public network. A public cloud is owned by an organization selling cloud services, and serves a diverse pool of clients.

For private clouds, the cloud infrastructure is operated exclusively for a single organization. A private cloud gives the organization exclusive access to and usage of the infrastructure and computational resources. It may be managed either by the organization or by a third party, and may be implemented at the organization's premise (i.e., *on-site private clouds*) or outsourced to a hosting company (i.e., *outsourced private clouds*).

Similar to private clouds, a community cloud may be managed by the organizations or by a third party, and may be implemented on customer premise (i.e., *on-site community cloud*) or outsourced to a hosting company (i.e., *outsourced community cloud*). However, a community cloud serves a set of organizations that have common security, privacy, and compliance considerations, rather than serving a single organization as does a private cloud.

A hybrid cloud is a composition of two or more clouds (private, community, or public) that remain unique entities but are bound together by standardized or proprietary technology that enables data and application portability. As discussed in this clause, both private clouds and community clouds can be either implemented on-site or outsourced to a third party. Therefore, each constituent cloud of a hybrid cloud can be one of the five variants.

3.3.2 Service Orchestration

Service orchestration refers to the arrangement, coordination, and management of cloud infrastructure to provide different cloud services to meet IT and business requirements. Figure 4 shows the general requirements and processes for cloud providers to build each of the three service models.

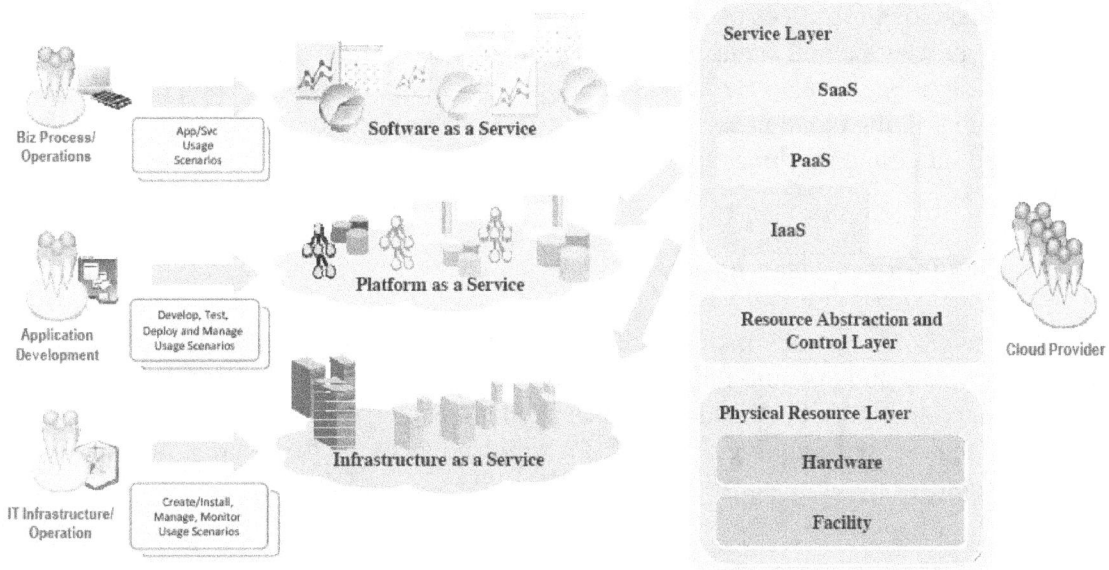

Figure 4 – Cloud Provider: Service Orchestration

A three-layered framework is identified for a generalized cloud environment in Figure 4. The top layer is the *service layer*, where a cloud provider defines and provisions each of the three service models. This is where cloud consumers consume cloud services through the respective cloud interfaces.

The middle layer is the *resource abstraction and control layer*. This layer contains the system components that a cloud provider uses to provide and manage access to the physical computing resources through software abstraction. The layer typically includes software elements such as hypervisors, virtual machines, virtual data storage, and other resource abstraction and management components needed to ensure efficient, secure, and reliable usage. While virtual machine technology is commonly used at this layer, other means of providing the necessary software abstractions are not precluded. This layer provides "cloud readiness" with the five characteristics defined in the NIST definition of cloud computing.

The lowest layer in the framework is the *physical resource layer*, which includes all the physical computing resources. This layer includes hardware resources, such as computers (CPU and memory), networks (routers, firewalls, switches, network links, and interfaces), storage components (hard disks), and other physical computing infrastructure elements. It also includes facilities resources, such as heating, ventilation, and air conditioning (HVAC), power, communications, and other aspects of the physical plant.

Note that in this framework, the horizontal positioning of layers implies a stack in which the upper layer has a dependency on the lower layer. The resource abstraction and control layer build virtual cloud resources on top of the underlying physical resource layer and support the service layer where cloud services interfaces are exposed. The three service models can be built either on top of one another (i.e., SaaS built upon PaaS and PaaS built upon IaaS) or directly upon the underlying cloud infrastructure. For example, a SaaS application can be implemented and hosted on virtual machines from IaaS or directly on top of cloud resources without using IaaS.

3.3.3 Cloud Service Management

Cloud Service Management includes all of the service-related functions that are necessary for the management and operation of those services required by or proposed to cloud consumers. As illustrated in Figure 5, cloud service management can be described from the perspective of *business support, provisioning and configuration,* and from the perspective of *portability and interoperability* requirements.

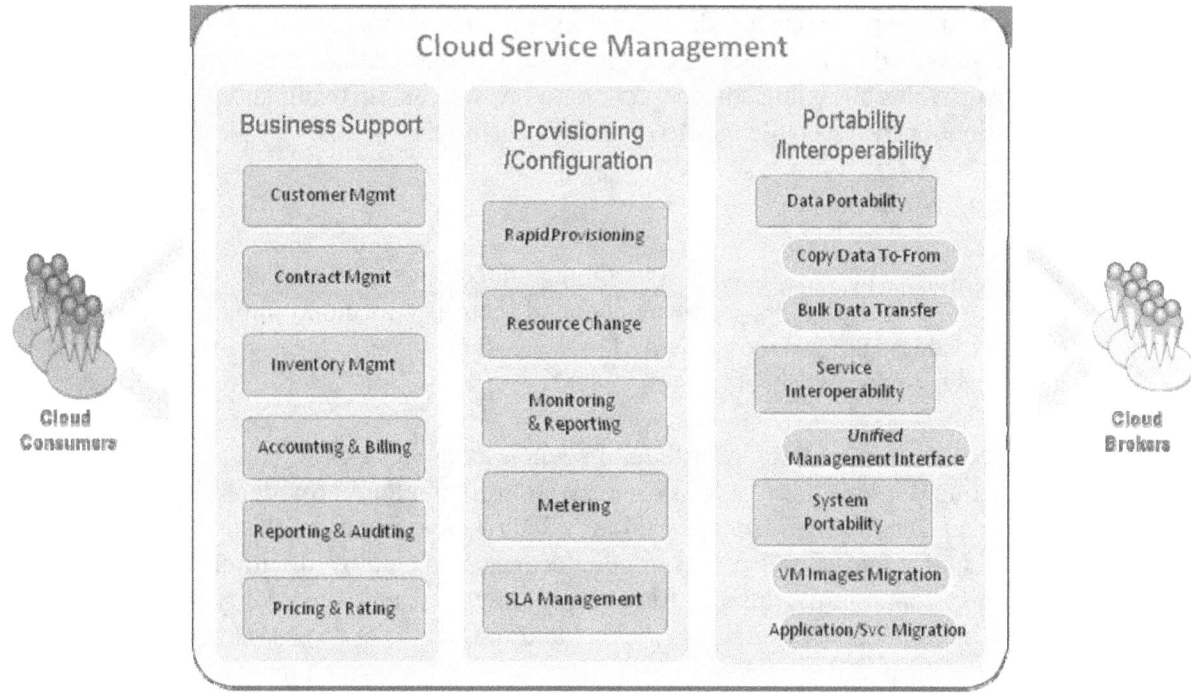

Figure 5 – Cloud Provider: Cloud Service Management

3.3.4 Security

"As the Federal Government moves to the cloud, it must be vigilant to ensure the security and proper management of government information to protect the privacy of citizens and national security." (*by Vivek Kundra, Federal Cloud Computing Strategy, Feb. 2011*.) It is critical to recognize that security is cross-cutting that spans across all layers of the reference model, ranges from physical security to application security, and in general, shares the responsibility between cloud provider and federal cloud consumer. For example, the protection of the physical resource layer (see Figure 4) requires physical security that denies unauthorized access to the building, facility, resource, or stored information. Cloud providers should ensure that the facility hosting cloud services is secure and that their staff has proper background checks. When data or application is moved to a cloud, it is important to ensure that the cloud offering satisfies the security requirements and enforces the compliance rules. An independent audit (see Clause 3.4) should be conducted to verify the compliance with regulation or security policy.

3.3.5 Privacy

Cloud providers should protect the assured, proper, and consistent collection, processing, communication, use and disposition of personal information (PI) and personally identifiable information (PII) in the cloud. According to the CIO Council, one of the federal government's key business imperatives is to ensure the privacy of the collected personally identifiable information. PII is the information that can be used to distinguish or trace an individual's identity, such as their name, social security number, biometric records, etc., alone, or when combined with other personal or identifying information that is linked or linkable to a specific individual, such as date and place of birth, mother's maiden name, etc. Though cloud computing provides a flexible solution for shared resources, software and information, it also poses additional privacy challenges to consumers using the clouds.

3.4 Cloud Auditor

A cloud auditor is a party that can conduct independent assessment of cloud services, information system operations, performance, and security of a cloud implementation. A cloud auditor can evaluate the services provided by a cloud provider in terms of security controls, privacy impact, performance, etc.

Auditing is especially important for federal agencies as "agencies should include a contractual clause enabling third parties to assess security controls of cloud providers" (*by Vivek Kundra, Federal Cloud Computing Strategy, February 2011*.). Security controls are the management, operational, and technical safeguards or countermeasures employed within an organizational information system to protect the confidentiality, integrity, and availability of the system and its information. For security auditing, a cloud auditor can make an assessment of the security controls in the information system to determine the extent to which the controls are implemented correctly, operating as intended, and producing the desired outcome with respect to the security requirements for the system. The security auditing should also include the verification of the compliance with regulation and security policy.

Federal agencies should be aware of the privacy concerns associated with the cloud computing environment where data are stored on a server that is not owned or controlled by the federal government. Privacy impact auditing can be conducted to measure how well the cloud system conforms to a set of established privacy criteria. A privacy impact audit can help federal agencies comply with applicable privacy laws and regulations governing an individual's privacy, and to ensure confidentiality, integrity, and availability of an individual's personal information at every stage of development and operation.

3.5 Cloud Broker

As cloud computing evolves, the integration of cloud services can be too complex for cloud consumers to manage. A cloud consumer may request cloud services from a cloud broker, instead of contacting a cloud provider directly. A cloud broker is an entity that manages the use, performance, and delivery of cloud services and negotiates relationships between cloud providers and cloud consumers.

In general, a cloud broker can provide services in three categories:

- *Service Intermediation*: A cloud broker enhances a given service by improving some specific capability and providing value-added services to cloud consumers. The improvement can be managing access to cloud services, identity management, performance reporting, enhanced security, etc.

- *Service Aggregation*: A cloud broker combines and integrates multiple services into one or more new services. The broker provides data integration and ensures the secure data movement between the cloud consumer and multiple cloud providers.

- *Service Arbitrage*: Service arbitrage is similar to service aggregation except that the services being aggregated are not fixed. Service arbitrage means a broker has the flexibility to choose services from multiple agencies. The cloud broker, for example, can use a credit-scoring service to measure and select an agency with the best score.

3.6 Cloud Carrier

A cloud carrier acts as an intermediary that provides connectivity and transport of cloud services between cloud consumers and cloud providers. Cloud carriers provide access to consumers through network, telecommunication, and other access devices. For example, cloud consumers can obtain cloud services through network access devices, such as computers, laptops, mobile phones, mobile IInternet devices (MIDs), etc. The distribution of cloud services is normally provided by network and telecommunication carriers or a *transport agent*, where a transport agent refers to a business organization that provides physical transport of storage media such as high-capacity hard drives. Note that a cloud provider will set up service level agreements (SLAs) with a cloud carrier to provide services consistent with the level of SLAs offered to cloud consumers, and may require the cloud carrier to provide dedicated and encrypted connections between cloud consumers and cloud providers.

4 Cloud Computing Use Cases

Cloud computing use cases describe the consumer requirements in using cloud computing service offerings. Analyzing business and technical cloud computing use cases and the applicable standards provides an intuitive, utility-centric perspective in surveying existing standardization efforts and identifying gaps. This clause leverages the business and technical use case outputs from other NIST Cloud Computing Program Working Groups and presents an analysis on how existing cloud-related standards fit the needs of USG cloud consumers and where the gaps for standardizations are.

4.1 Business Use Cases

The Target Business Use Case Working Group has produced a template for documenting specific use cases. This template includes a section titled "Concept of Operations" in which "Current System" and "Desired Cloud Implementation" states are described. The template also gathers information about integration with other systems, security requirements, and both local and remote network access considerations. A set of business use cases is being drafted describing candidate USG agency cloud deployments. The stories captured in these business use cases help to identify business drivers behind the adoption of cloud computing in USG agencies, provide background information on the relevant usage context, and expose general agency consumer concerns and issues through specific scenarios. These use cases thus help us to document key technical requirements for USG cloud-related standards in the areas of security, interoperability and portability as required for the formulation of this roadmap.

The "Cloud First" business use case called out by the Federal CIO is a more general expansion of this analysis to multiple interacting current systems and cloud implementations. This expansion is to support evolving business processes as cloud deployments are implemented. It requires interoperability and portability across multiple cloud deployments and enterprise systems.

4.2 Technical Use Cases

The SAJACC Working Group has produced a set of preliminary use cases developed for the SAJACC project for the first pass through the SAJACC process. Through a series of open workshops, and through public comment and feedback, NIST will continue to refine these use cases and add new use cases as appropriate. These use cases are technical in nature, capturing the more generic and cross-cutting technical requirements of cloud consumers. They are descriptions of how groups of users and their resources may interact with one or more cloud computing systems to achieve specific goals, such as "how to copy data objects into a cloud."

There is a natural mapping from the high-level business use cases to the SAJACC technical use cases, where the business operational stories of specific agency consumers will imply specific technical requirements expressed in SAJACC technical use cases. For example, the business use case of an agency consumer's move of its virtualized computing infrastructure to an IaaS cloud vendor implies the technical requirement of "*Virtual Machine (VM) control: manage virtual machine instance state"* to be met. The rest of this clause drives through the high-level business use cases to the general technical requirements expressed and analyzes where cloud standards help address these requirements.

4.3 Deployment Scenario Perspective

The "Cloud First" business use case requires more complex interactions between USG agency cloud consumer and cloud providers. There are three main groups of interaction scenarios:

The figure below illustrates the different generic scenarios.

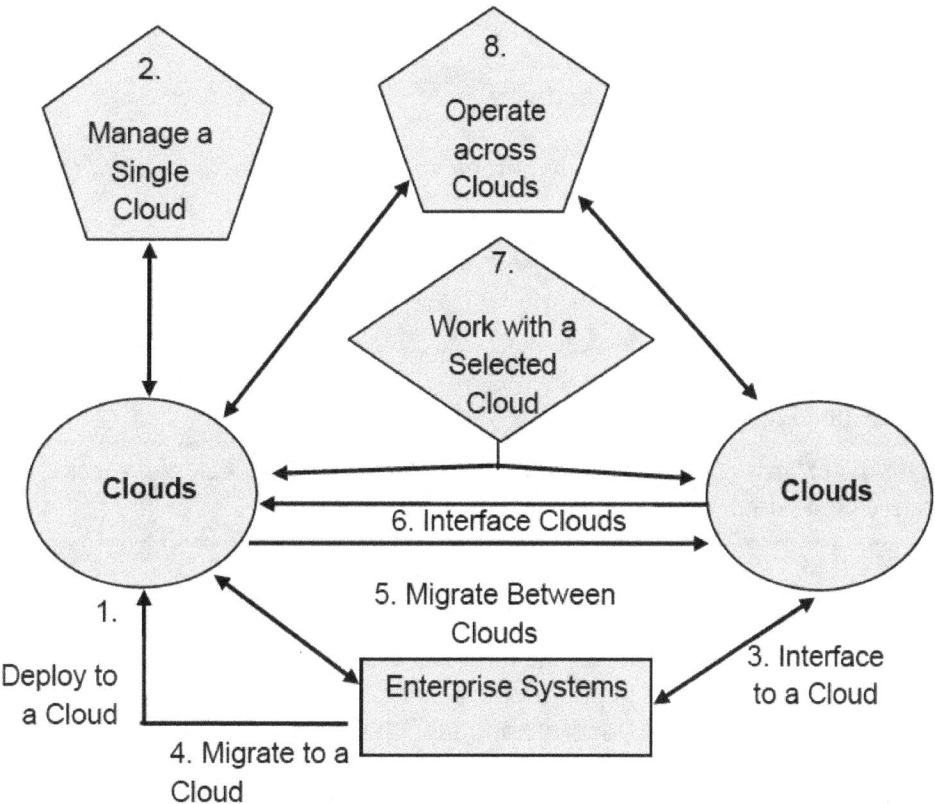

Figure 6 – High-Level Generic Scenarios

Single Cloud
    Scenario 1: Deployment on a single cloud
    Scenario 2: Manage resources on a single cloud
    Scenario 3: Interface enterprise systems to a single cloud
    Scenario 4: Enterprise systems migrated or replaced on a single cloud

Multiple Clouds (serially, one at a time)
    Scenario 5: Migration between clouds
    Scenario 6: Interface across multiple clouds
    Scenario 7: Work with a selected cloud

Multiple Clouds – (simultaneously, more than one at a time)
    Scenario 8: Operate across multiple clouds

These technical use cases must also be analyzed in the context of their deployment models and the resultant way cloud actors must interact. These considerations identify two fundamental dimensions to the spectrum of cloud computing use cases:

- Centralized vs. Distributed, and
- Within vs. Crossing Trust Boundaries

These deployment cases will drive the requirements for cloud standards. They can be identified through the following matrix:

	a.) Within Trust Boundary	b.) Crossing Trust Boundary
1.) Centralized i.e., one administrative cloud domain	Deployment Case 1A	Deployment Case 1B
2.) Distributed, i.e., crossing administrative cloud domains	Deployment Case 2A	Deployment Case 2B

Table 3 – Deployment Cases for High-Level Scenarios

Deployment Case 1: In the centralized deployment cases, there is one cloud provider under consideration at a time. Each cloud provider may service multiple cloud consumers. Each cloud consumer has a simple client-provider interaction with the provider.

Deployment Case 1A: This deployment case is typically a *private cloud* within a single administrative domain and trust boundary wherein policy and governance can be enforced by nontechnical means. Use cases within this deployment case may require standards to support the following basic technical requirements:

- Simple, consumer-provider authentication;
- VM management;
- Storage management;
- SLAs and performance/energy monitoring;
- Service discovery;
- Workflow management;
- Auditing; and
- Virtual organizations in support of community cloud use cases.

Deployment Case 1B: This deployment case is typically (commercial) *public cloud* within a single administrative domain but is outside of any trust boundary that a client could use to enforce policy and governance. Clients must rely on the cloud provider to enforce policy and governance through technical means that are "baked into" the infrastructure. Use cases within this deployment case may require standards to support the following additional technical requirements:

- SLAs in support of governance requirements, e.g., national or regional regulatory compliance;
- Stronger authentication mechanisms, e.g., Public Key Infrastructure (PKI) Certificates, etc.;
- Certification of VM isolation through hardware and hypervisor support;
- Certification of storage isolation through hardware support; and
- Data encryption,

Deployment Case 2: In the distributed deployment cases, a single cloud consumer has an application that may be distributed across two or more cloud providers and administrative domains simultaneously. While the cloud consumer may have simple consumer-provider interactions with their application and the providers, more complicated *Peer-to-Peer ("P2P")* interactions may be required -- between both the consumer and provider and also between the providers themselves.

Deployment Case 2A: This deployment case is typically a federated cloud of two or more administrative cloud domains, but where the cloud providers can agree "out of band" how to mutually enforce policy and governance -- essentially establishing a common trust boundary. Use cases within this deployment case may require standards to support the following basic technical requirements:

- P2P service discovery;
- P2P SLA and performance monitoring;
- P2P workflow management;
- P2P auditing;
- P2P security mechanisms for authentication, authorization; and
- P2P virtual organization management.

Deployment Case 2b: This deployment case is typically a *hybrid cloud* where apps cross a private-public trust boundary, or even span *multiple public clouds*, where both administrative domains and trust boundaries are crossed. Consumers must rely on the cloud provider to enforce policy and governance through technical means that are "baked into" the infrastructure. Apps and services may be distributed and need to operate in a P2P manner. Use cases within this deployment case will require all the standards of the other deployment cases, in addition to the following more extensive technical requirements:

- P2P SLAs in support of governance requirements.

The use cases presented in this clause will be analyzed with regards to their possible *deployment scenarios* to determine their requirements for standards. This analysis will be subsequently used to evaluate the likelihood of each of these deployment cases. Clearly the expected deployment of these use cases across the different deployment cases will not be uniform. This non-uniformity will assist in producing a *prioritized roadmap* for cloud standards. Likewise, in reviewing existing standards, these use cases – in conjunction with their possible deployment cases – will be used to identify and prioritize *gaps* in available standards.

Based on this analysis, we note that Scenarios 1 through 4 could, in fact, be deployed on either a private cloud or a public cloud. Hence, the different standards noted in deployment cases 1A and 1B will be required. Scenarios 5, 6, and 7 all involve the notion of the serial use of multiple clouds. Presumably these different clouds, used serially, could be either private or public. Hence, deployment cases 1A and 1B would also apply, but there are additional requirements to achieve portability, e.g., Application Programming Interface (API) commonality. Finally, Scenario 8 could involve a federated/community cloud or a hybrid cloud. Hence, deployment cases 2A and 2B would apply here.

To summarize the detailed technical use cases for this analysis, the following areas of technical requirements are common across all scenarios:

1. Creating, accessing, updating, deleting data objects in clouds;
2. Moving VMs and virtual appliances between clouds;
3. Selecting the best IaaS vendor for private externally hosted cloud;
4. Tools for monitoring and managing multiple clouds;
5. Migrating data between clouds;
6. Single sign on access to multiple clouds;
7. Orchestrated processes across clouds;
8. Discovering cloud resources;
9. Evaluating SLAs and penalties; and
10. Auditing clouds.

5 Cloud Computing Standards

Standards are already available in support of many of the functions and requirements for cloud computing described in Clauses 3 and 4. While many of these standards were developed in support of pre-cloud computing technologies, such as those designed for web services and the Internet, they also support the functions and requirements of cloud computing. Other standards are now being developed in specific support of cloud computing functions and requirements, such as virtualization.

To assess the state of standardization in support of cloud computing, the NIST Cloud Computing Standards Roadmap Working Group has compiled an Inventory of Standards Relevant to Cloud Computing http://collaborate.nist.gov/twiki-cloud-computing/bin/view/CloudComputing/StandardsInventory. This inventory is being maintained and will be used to update this document as necessary.

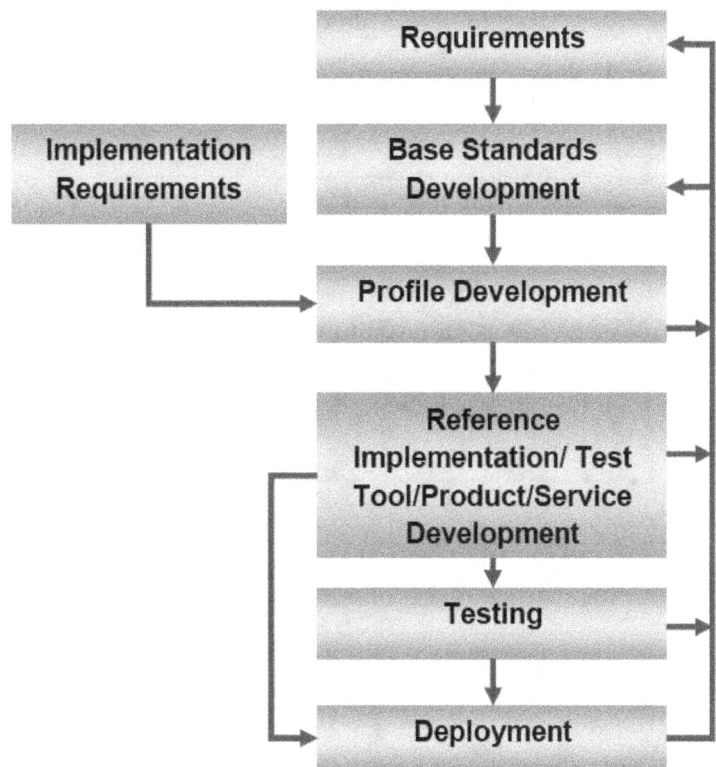

Figure 7 – IT Standards Life Cycle

5.1 Information and Communication Technologies (IT) Standards Life Cycle

Figure 7 is a high-level conceptualization of how IT standards are developed and standards-based IT products, processes and services are deployed. This figure is not meant to imply that these processes occur sequentially. Many of the processes illustrated can and should be done somewhat concurrently. Some of these processes (eg, reference implementations / product / process / service / test tools development; testing; deployment) occur outside of the SDO process. These processes provide input and feedback to improve the standards, profiles, test tools, etc.

5.2 Categorizing the Status of Standards

Innovation in IT means that IT standards are constantly being developed, approved, and maintained. Revisions to previous editions of standards may or may not be backward-compatible. Table 4 is intended to provide an indication of the maturity level of a standard. Some SDOs require two or more implementations before final approval of a standard. Such implementations may or may not be commercial products or services. In other cases, an SDO may be developing a standard while conforming commercial products or services are already being sold.

Maturity Level	Definition
No Standard	SDOs have not initiated any standard development projects.
Under Development	SDOs have initiated standard development projects. Open source projects have been initiated.
Approved Standard	SDO-approved standard is available to public. Some SDOs require multiple implementations before final designation as a "standard."
Reference Implementation	Reference implementation is available
Testing	Test tools are available. Testing and test reports are available.
Products/Services	Standards-based products/services are available.
Market Acceptance	Widespread use by many groups. De facto or de jure market acceptance of standards-based products/services.
Sunset	Newer standards (revisions or replacements) are under development.

Table 4 – Standards Maturity Model

5.3 Cloud Computing Standards for Interoperability

As it would be expected there are a broad range of capabilities and functions available in the various cloud provider interfaces currently available. This may indicate that we are still in the early days of cloud computing and consolidation has not yet occurred. While standardization of cloud interfaces are maturing, commonalities among provider interfaces can help us understand the key interoperability requirements and features.

The interfaces that are presented to cloud users can be broken down into two major categories, with interoperability determined separately for each category. As show in the diagrams below, each type of cloud offering presents an interface of each category.

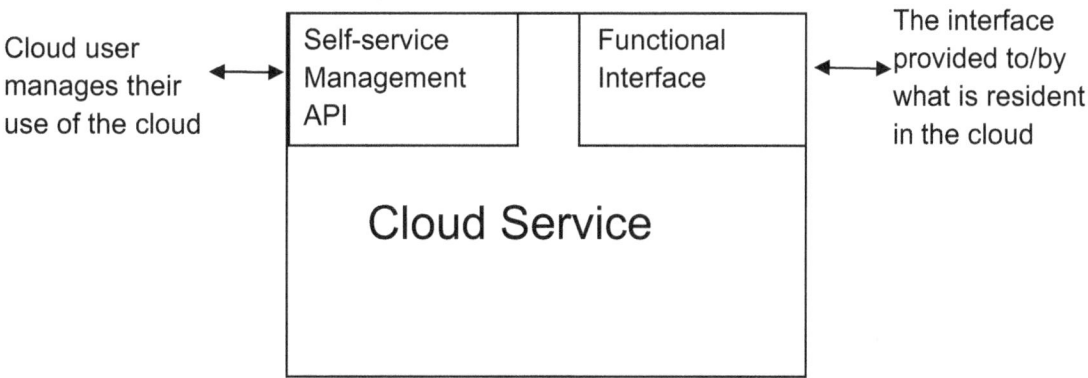

Figure 8 – Cloud Service Presents an Interface to Each Category

The interface that is presented to (or by) the contents of the cloud encompasses the primary *function* of the cloud service. This is distinct from the interface that is used to *manage* the use of the cloud service. For an Infrastructure as a Service cloud offering, as shown in the diagram below, the **Functional Interface** is a virtualized Central Processing Unit (CPU), Memory and Input/Output (I/O) space typically used by an operating system (and the stack of software running in that operating system [OS[instance).

The cloud user utilizes the **Management Interface** to control their use of the cloud service by starting, stopping, and manipulating virtual machine images and associated resources. It should be clear from this that the Functional Interface for an IaaS cloud is very much tied to the architecture of the CPU that is being virtualized. This is not a cloud-specific interface and no effort is being put into a de jure standard for this interface since de facto CPU architectures are the norm.

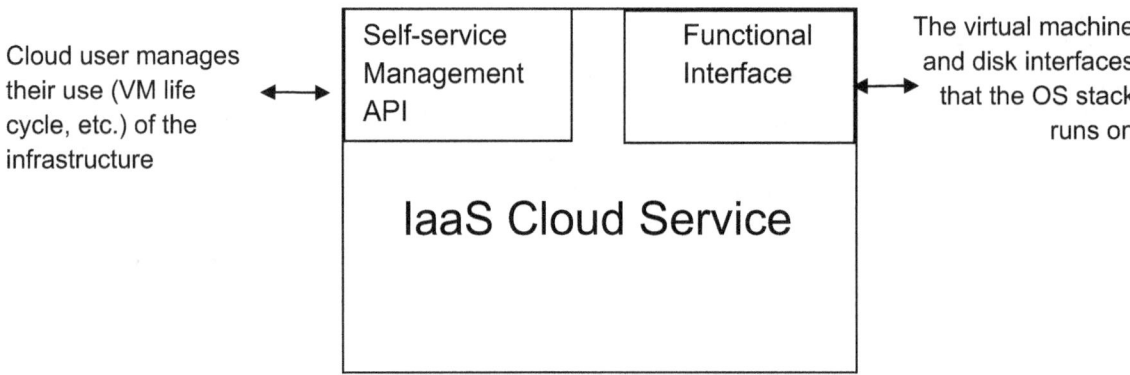

Figure 9 – IaaS Interface

The self-service IaaS management interface, however, is a candidate for interoperability standardization, and there are several efforts in this space. The Open Cloud Computing Interface (OCCI) interface from the Open Grid Forum is an example of a standard IaaS resource management interface. The Cloud Data Management Interface (CDMI) standard is an example of both storage management interface as well as a storage functional interface. There is a rapid proliferation of various proprietary interfaces as well as all competing to become a de facto means of interoperability.

For PaaS, as shown below, again we see the differentiation needed between these two categories of interfaces.

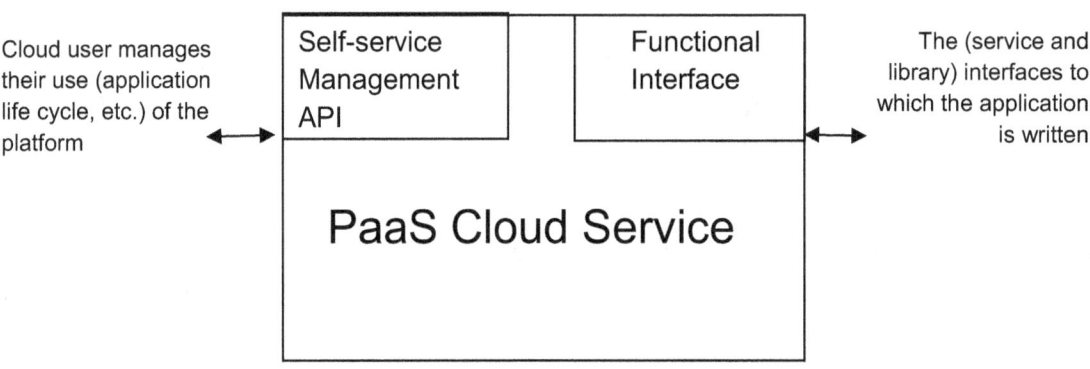

Figure 10 – PaaS Interface

The functional interface of a PaaS offering is a runtime environment with a set of libraries and components to which the application is written. This could be offered in different languages and may or may not take advantage of existing application platforms standards such as those found in J2EE or .NET. The Management Interface of a PaaS offering, however, may be very similar to the Management Interface of an IaaS offering. Instead of the life cycle of virtual machines and their resources, the PaaS self-service interface is concerned with the life cycle of applications and the platform resources they depend on. In addition, instead of being metered and billed on the basis of virtual hardware resources, the interface typically exposes metrics for platform service and runtime container usage. Interoperability of Paas self-service management interfaces can be achieved separately from the interoperability of the PaaS functional interfaces, although there seem to be very few efforts concentrating on PaaS management interfaces today.

For Software as a Service offering, as shown below, the functional interface is the same as the application interface of the software itself. In the case where a SaaS application is consumed through a Web browser, there may be many standards that are used to achieve interoperability between what is essentially a Web server and the user's browser, such as IP (v4, v6), TCP, HTTP. SSL/TLS, HTML, XML, REST, Atom, AtomPub, RSS, and JavaScript/JSON. None of these Web standards are cloud-specific, and these same standards are being used in the many Web browser-based management interfaces.

In the case where a SaaS application is consumed by another system as a service, cloud or otherwise, there are various standards as to both data content and interfaces. Most important for interoperability are canonical data content formats, typically today expressed using XML standards. Such standard canonical formats include "nouns," i.e., the data objects being acted on, but also (implicitly or explicitly) the "verbs," i.e., the actions that a receiving service may or should take on such a data object (e.g., Sync, Process, Get, Show, etc.). While "verbs" may be somewhat generic, such canonical formats are in general specific to a particular domain. Various standards exist corresponding to different application domains (e.g., OAGi BODs for business documents or ODF and OOXML for office productivity documents). Also important is the stack of interoperability standards for interfaces, packaging, and transport such as SOAP, WS-* and ebXML.

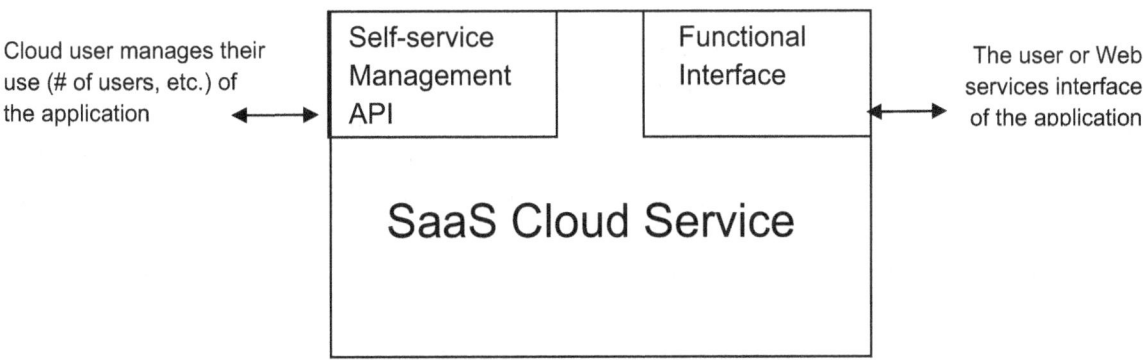

Figure 11 – SaaS Interface

The self-service management interface of a SaaS offering is typically concerned, not with life cycle, but with the administration and customization of application functionality for each user of the offering. Through this interface, for example, additional users can be added (along with their credentials and permissions), additional features can be ordered for each user (usually in packaged sets), and an accounting of each user's consumption of the offering is available. Interoperability of a SaaS management interface may be best achieved by focusing initially on Web service interfaces for common operations, such as those around identity management.

Most of these interfaces will be tested and analyzed by NIST to validate its capabilities against the list of cloud computing use cases. At the same time, work is continuing in the SDOs to further the interests of cloud computing interoperability – including the maintenance of standards to reflect implementation experience, development of new standards for agreed-upon functions and/or protocols, and the profiling of existing standards.

5.4 Cloud Computing Standards for Portability

The rapid adoption of virtual infrastructure has popularized the practice of packaging, transporting and deploying pre-configured and ready-to-run systems, including all needed applications and the operating systems into virtual machines. The development of a standard, portable meta-data model for the distribution of virtual machines to and between virtualization and cloud platforms will enable the portability of such packaged workloads on any cloud computing platform. Some cloud workload formats contain a single VM only; modern enterprise applications are often constructed using a multiple tiered model, where each tier contains one or more machines. A single VM model is thus not sufficient to distribute a complete multitiered system. In addition, complex applications require install-time customization of networks and other customer-specific properties. Furthermore, a virtual machine image is packaged in a run-time format with hard disk images and configuration data suitable for a particular hypervisor. Run-time formats are optimized for execution and not for distribution. For efficient software distribution, a number of additional features become critical, including platform independence, compression, verification, signing, versioning, and software licensing management, temporal synchronization of state metadata snapshots and federated identification by organization and devices with organizations.

Over the last year, much progress has been made on new standards in this area. Open Virtualization Format (OVF) from the Distributed Management Task Force (DMTF), for example, was developed to address portability concerns between various virtualization platforms. It consists of metadata about a virtual machine image or groups of images that can be deployed as a unit. It provides an easy way to package and deploy services as either a virtual appliance or used within an enterprise to prepackage known configurations of a virtual machine image or images. It may contain information regarding the number of CPUs, memory required to run effectively, and network configuration information. It also can contain digital signatures to ensure the integrity of the machine images being deployed along with licensing information in the form of a machine-readable EULA (End User License Agreement) so that it can be understood before the image(s) is deployed.

A future direction of workloads data and metadata standardization is to help improve the automation of inter-cloud workload deployment. Concepts such as standardized SLAs, sophisticated inter virtual machine network configuration and switching information, and software license information regarding all of the various components that make up the workload are possibilities.

Another aspect of portability in the cloud environment is that of storage and data (including metadata) portability between clouds, for example, between storage cloud services and between compatible application services in SaaS and PaaS layers.

Cloud storage services may be seen as a special class of application service, where the storage metadata (as distinct from the stored data content) is the application data that a receiving cloud system must be able to process. For cloud storage services, as much of the actual data movement needs to be done in bulk moves of massive numbers of objects, retaining the data organization (into containers, for example) and retaining the associated metadata are main portability requirements.

Data portability between cloud application services requires standard formats and protocols for the data to be moved. The canonical data formats commonly involved in portability scenarios may be focused on widely used application categories, for example email or office productivity, or on specific formats used by particular domains of use, for example science or medical domains. Popular methods for interchange of data in clouds generally leverage representations in either JSON or XML formats, and are often customized to particular fields of use through specialized standards.

5.4.1 Workload Portability in the Cloud

Workloads and data need to be able to move around. Cloud platforms should make it easy and efficient to securely move customer applications on and off, and data in and out, of their infrastructure. There should be a secure migration path to cloud computing that preserves existing investments in technologies which are appropriate to the cloud, and that enables the coexistence and interoperability of on-premises software and cloud services. Application and data portability, in particular, are key considerations and should prevent vendor lock-in, whether moving to the cloud in the first place or moving from one cloud to another.

Application and data portability is a key requirement, whether moving to the cloud in the first place or moving from one cloud to another. Organizations that have virtualized their datacenters have already taken the first step to the cloud. Packaging an operating system, application, and data in a VM reduces disruption from different hardware options. However the provider's implementation of all the details of VM packing and management may be different from the consumer's. As a result:

- The application appliance may not be accepted by the destination cloud.
- The application may not start.
- The application may execute but fail to behave as expected.
- Performance may be poor.
- Bulk data may move to the cloud incorrectly.
- VMs may not respond to the management commands.

The IT staff's first cloud encounter of such portability challenges is involved in using IaaS (where the VMs are executed) and PaaS (where data, identity, and access are managed).

Standards are key to achieving portability. Building on existing standards and specifications that are known to work and already in widespread use (and documenting how the standards are implemented) allows developers to continue to use their chosen development languages and tools as they build for cloud environments. This keeps migration costs and risks low by enabling organizations to leverage their IT staff's current skills, and by providing a secure migration path that preserves existing investments. Examples of languages, tools, and standards that are common in the cloud include programming languages such as Java, C#, PHP, Python and Ruby; Internet protocols for service access such as REST, SOAP, and XML; federated identity standards for service authentication such as SAML and Oauth; and standards for managing virtualized environments.

Standards continue to rapidly evolve in step with technology. Hence, cloud standards may be at different stages of maturity and levels of acceptance. OVF, for example, is an open standard for packaging and distributing virtual appliances. Originally offered as a proprietary format to the DMTF, OVF was first published in March 2009, and subsequently adopted in August 2010 as a national standard by the American National Standards Institute (ANSI). When a provider claims conformance with OVF or any other standard, it should cite the specific version and publish implementation, errata, and testing notes. This will provide the transparency necessary for informed consumer choice, as well as to ensure reasonably seamless technical interoperability between on-premises and cloud virtualized environments.

5.4.2   Data Portability in the Cloud

Many people are focused on the need for cloud portability as the means to prevent being locked into any particular cloud or provider. *Portability* is generally the ability to move applications and data from one computing environment to another. But there are differences between *application* and *data* portability.

With regard to applications, virtualization has greatly improved the portability of server-based workloads. Early on (and still to some extent), some cloud providers used proprietary virtual machine images that are difficult to map to enterprise networks and require transformations in order to port. Now, with the maturing and widespread adoption of the virtualization management standards (mentioned above), open standards are in place to facilitate VM portability among conformant cloud providers.

Data portability is more complex and more fundamental to the notion of portability. It puts the ultimate control over data in the hands of the owner of that data, not the Web application that uses it or the service provider that hosts the application. Cloud consumers need to maintain control of their data. Moving data from one cloud provider to another includes the need to securely delete the old storage space.

Complexities in data portability stem from the fact that applications process different volumes, kinds, and forms of data, and this data may flow throughout the entire system. For example, a financial application might use a petabyte of data, but that data might be securely housed in a single cloud database, making it relatively easy to port. On the other hand, a customer relationship management (CRM) application running in the cloud might process only a terabyte of data but which is shared among thousands of users; moving the CRM application – and all its distributed data – from one cloud to another would be more challenging. The key to data portability is that the user's data and metadata (i.e., data about the data) are available in a well-documented and well-tested format available to all for use on other platforms.

It can be claimed that as long as users' data is not locked in, thanks to well-documented and easily accessible interoperable interfaces, the user is not locked in, and that moving to another cloud provider is just a matter of enduring a switching cost. Such cost can be lowered by using best practices such as choosing cloud providers who support a wide range of programming languages and application runtimes and middleware, as well as a variety of cloud deployment models independent of other choices that the user may have made.

The data porting process is not complete, however, until the data is removed or erased from the old cloud provider. Sometimes called the "right to be forgotten," the consumer's ability to delete data is as essential to a user's control over that data as the ability to retrieve it. This is a hot topic for debate currently as governments consider regulations and the industry works on technical solutions to address this issue in a standardized way. Protocols for transport of data are receiving attention from the standards community.

5.5 Cloud Computing Standards for Security

The three cybersecurity objectives, ensuring the confidentiality, integrity, and availability of information and information systems, are particularly relevant as these are the high-priority concerns and perceived risks related to cloud computing. Cloud computing implementations are subject to local physical threats as well as remote, external threats. Consistent with other applications of IT, the threat sources include accidents, natural disasters and external loss of service, hostile governments, criminal organizations, terrorist groups, and intentional and unintentional introduction of vulnerabilities through internal and external authorized and unauthorized human and system access, including but not limited to employees and intruders. The characteristics of cloud computing, significantly multi-tenancy and the implications of the three service models and four deployment models, heighten the need to consider data and systems protection in the context of logical as well as physical boundaries.

Possible types of attacks against cloud computing services include the following:

- Compromises to the confidentiality and integrity of data in transit to and from a cloud provider;

- Attacks which take advantage of the homogeneity and power of cloud computing environments to rapidly scale and increase the magnitude of the attack;

- Unauthorized access by a consumer (through improper authentication or authorization, or vulnerabilities introduced during maintenance) to software, data, and resources in use by an authorized cloud service consumer;

- Increased levels of network-based attacks, such as denial of service attacks, which exploit software not designed for an Internet threat model and vulnerabilities in resources which were formerly accessed through private networks;

- Limited ability to encrypt data at rest in a multi-tenancy environment;

- Portability constraints resulting from nonstandard application programming interfaces (APIs) which make it difficult for a cloud consumer to change to a new cloud service provider when availability requirements are not met;

- Attacks which exploit the physical abstraction of cloud resources and exploit a lack of transparency in audit procedures or records;

- Attacks that take advantage of virtual machines that have not recently been patched; and

- Attacks which exploit inconsistencies in global privacy policies and regulations.

Major security objectives for a cloud computing implementation include the following:

- Protect customer data from unauthorized access, disclosure, modification or monitoring. This includes supporting identity management such that the customer has the capability to enforce identity and access control policies on authorized users accessing cloud services. This includes the ability of a customer to make access to its data selectively available to other users.

- Protect from supply chain threats. This includes ensuring the trustworthiness and reliability of the service provider as well as the trustworthiness of the hardware and software used.

- Prevent unauthorized access to cloud computing infrastructure resources. This includes implementing security domains that have logical separation between computing resources (e.g. logical separation of customer workloads running on the same physical server by VM monitors [hypervisors] in a multitenant environment) and using secure-by-default configurations.

- Design Web applications deployed in a cloud for an IInternet threat model and embedding security into the software development process.

- Protect Internet browsers from attacks to mitigate end-user security vulnerabilities. This includes taking measures to protect Internet-connected personal computing devices by applying security software, personal firewalls, and patch maintenance.

- Deploy access control and intrusion detection technologies at the cloud provider, and conduct an independent assessment to verify that they are in place. This includes (but does not rely on) traditional perimeter security measures in combination with the domain security model. Traditional perimeter security includes restricting physical access to network and devices; protecting individual components from exploitation through security patch deployment; setting as default most secure configurations; disabling all unused ports and services; using role-based access control; monitoring audit trails; minimizing the use of privilege; using antivirus software; and encrypting communications.

- Define trust boundaries between service provider(s) and consumers to ensure that the responsibility for providing security is clear.

- Support portability such that the customer can take action to change cloud service providers when needed to satisfy availability, confidentiality, and integrity requirements. This includes the ability to close an account on a particular date and time, and to copy data from one service provider to another.

6 Cloud Computing Standards Mapping and Gap Analysis

One approach to relevant cloud standards mapping and gap analysis is to map relevant cloud standards using the conceptual model and the cloud computing taxonomy from the NIST Cloud Computing Reference Architecture and Taxonomy Working Group. As presented in Figure 12: cloud computing conceptual model, the cloud computing conceptual model is depicted as an integrated diagram of system, organizational, and process components. The cloud computing taxonomy produced by the same working group has provided further categorizations for the security, interoperability, and portability aspects for cloud computing. While many standards are generally relevant to these cloud computing areas, the following sections will map those specifically relevant cloud standards and capture their standard maturity status in a tabular format. The online cloud standards inventory (as described in Clause 5) will be the place to maintain and track other more general relevant standards. Some standards may apply to more than one category from the cloud taxonomy and therefore may be listed more than once.

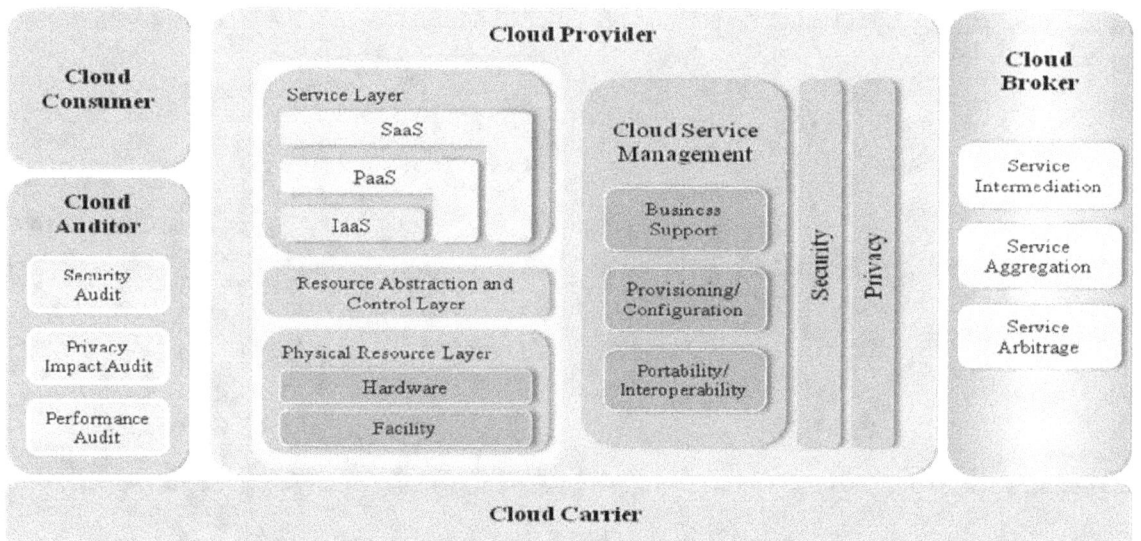

Figure 12 – The Combined Conceptual Reference Diagram

6.1 Security Standards Mapping

The table below maps standards to the security categories in the NIST Cloud Computing Taxonomy and gives their status (ref: Table 4, Standards Maturity Model). Some of the listed standards apply to more than one category and are therefore listed more than once.

Categorization	Available Standards and SDO	Status
Authentication & Authorization	RFC 5246: Secure Sockets Layer (SSL)/ Transport Layer Security (TLS); IETF	Approved Standard Market Acceptance
	RFC 3820: X.509 Public Key Infrastructure (PKI) Proxy Certificate Profile; IETF	Approved Standard Market Acceptance
	RFC5280:Internet X.509 Public Key Infrastructure Certificate and Certificate Revocation List (CRL) Profile; IETF	Approved Standard Market Acceptance
	X.509	ISO/IEC 9594-8: Information technology – Open systems interconnection – The Directory: Public-key and attribute certificate frameworks, ITU-T	Approved Standard Market Acceptance
	RFC 5849: Oauth (Open Authorization Protocol); IETF	Approved Standard Market Acceptance
	OpenID Authentication; OpenID	Approved Standard Market Acceptance
	eXtensible Access Control Markup Language (XACML); OASIS	Approved Standard Market Acceptance
	Security Assertion Markup Language (SAML); OASIS	Approved Standard Market Acceptance
	FIPS 181: Automated Password Generator; NIST	Approved Standard Market Acceptance
	FIPS 190: Guideline for the Use of Advanced Authentication Technology Alternatives; NIST	Approved Standard Market Acceptance
	FIPS 196: Entity Authentication Using Public Key Cryptography; NIST	Approved Standard Market Acceptance
Confidentiality	RFC 5246: Secure Sockets Layer (SSL)/ Transport Layer Security (TLS); IETF	Approved Standard Market Acceptance
	Key Management Interoperability Protocol (KMIP); OASIS	Approved Standard Market Acceptance
	XML Encryption Syntax and Processing; W3C	Approved Standard Market Acceptance
	FIPS 140-2: Security Requirements for Cryptographic Modules; NIST	Approved Standard Market Acceptance
	FIPS 185: Escrowed Encryption Standard (EES); NIST	Approved Standard Market Acceptance
	FIPS 197: Advanced Encryption Standard (AES); NIST	Approved Standard Market Acceptance
	FIPS 188: Standard Security Label for Information Transfer; NIST	Approved Standard Market Acceptance
Integrity	XML signature (XMLDSig); W3C	Approved Standard Market Acceptance
	FIPS 180-3: Secure Hash Standard (SHS); NIST	Approved Standard Market Acceptance
	FIPS 186-3: Digital Signature Standard (DSS); NIST	Approved Standard Market Acceptance
	FIPS 198-1: The Keyed-Hash Message Authentication Code (HMAC); NIST	Approved Standard Market Acceptance

Categorization	Available Standards and SDO	Status
Identity Management	Service Provisioning Markup Language (SPML); WS-Federation and WS-Trust	Approved Standard
	X.idmcc – Requirement of IdM in Cloud Computing, ITU-T	Under Development
	Security Assertion Markup Language (SAML); OASIS	Approved Standard Market Acceptance
	OpenID Authentication, OpenID Foundation	Approved Standard Market Acceptance
	FIPS 201-1: Personal Identity Verification (PIV) of Federal Employees and Contractors, NIST	Approved Standard Market Acceptance
Security Monitoring & Incident Response	NIST SP 800-126: Security Content Automation Protocol (SCAP), NIST	Approved Standard Market Acceptance
	NIST SP 800-61 Computer Security Incident Handling Guide, NIST	Approved Standard
	X.1500 Cybersecurity information exchange techniques, ITU-T	Approved Standard Market Acceptance
	X.1520: Common vulnerabilities and exposures; ITU-T	Approved Standard
	X.1521; Common Vulnerability Scoring System; ITU-T	Approved Standard
	PCI Data Security Standard; PCI	Approved Standard Market Acceptance
	FIPS 191: Guideline for the Analysis of Local Area Network Security; NIST	Approved Standard Market Acceptance
Security Policy Mgmt	eXtensible Access Control Markup Language (XACML); OASIS	Approved Standard Market Acceptance
	FIPS 199: Standards for Security Categorization of Federal Information and Information Systems; NIST	Approved Standard Market Acceptance
	FIPS 200: Minimum Security Requirements for Federal Information and Information Systems; NIST	Approved Standard Market Acceptance
Availability	ISO/PAS 22399:2007 Guidelines for incident preparedness and operational continuity management, ISO	Market Acceptance

Table 5 – Security: Categorization

6.2 Interoperability Standards Mapping

As discussed in Clause 5.3, the interoperability of cloud services can be categorized by the management and functional interfaces of the cloud services. Many existing IT standards contribute to the interoperability between cloud consumer applications and cloud service, and between cloud services themselves. There are standardization efforts that are specifically initiated to address the interoperability issues in the cloud. These cloud specific standards are listed in Table 6 below.

Categorization	Available Standards and SDO	Status
Service Interoperability	Open Cloud Computing Interface (OCCI); Open Grid Forum	Approved Standard
	Cloud Data Management Interface (CDMI); Storage Networking Industry Association, SNIA	Approved Standard
	IEEE P2301, Draft Guide for Cloud Portability and Interoperability Profiles (CPIP), IEEE	Under Development
	IEEE P2302, Draft Standard for Intercloud Interoperability and Federation (SIIF), IEEE	Under Development

Table 6 – Interoperability: Categorization

6.3 Portability Standards Mapping

As discussed in Clause 5.4, portability issues in the cloud include workload and data portability. While some of the cloud workload portability issues are new, a lot of existing data and metadata standards have been developed before the cloud era. The following table focuses on cloud-specific portability standards.

Categorization	Available Standards and SDO	Status
Data Portability	Cloud Data Management Interface (CDMI); SNIA	Approved Standard
System Portability	Open Virtualization Format (OVF); DMTF	Approved Standard Market Acceptance
	IEEE P2301, Draft Guide for Cloud Portability and Interoperability Profiles (CPIP), IEEE	Under Development

Table 7 – Portability: Categorization

6.4 Use Case Analysis

There are several facets of cloud service interfaces that are candidates for standardization including:

- Management APIs;
- Data Exchange Formats;
- Federated Identity and Security Policy APIs;
- Resource Descriptions; and
- Data Storage APIs.

With these candidate areas in mind, the following business use cases can be analyzed with regard to their possible deployment modes (as discussed in Clause 4.3) to identify required standards. This analysis, in conjunction with the NIST Cloud Standards Inventory, enables the availability of relevant existing and emerging standards to be evaluated. Where no suitable standards of any kind exist, this is a gap. As part of this use case analysis, the priority of the standards or requirements in question is also identified.

6.4.1 Use Case: Creating, accessing, updating, deleting data objects in clouds

Benefits: Cross-cloud applications
Deployment Mode Considerations: Basic Create-Read-Update-Delete (CRUD) operations on data objects will primarily be done between a single client and provider, and should observe any required standards for authentication and authorization.
Standardizations Needed: Standard interfaces to metadata and data objects
Possible Standards: CDMI from SNIA

6.4.2 Use Case: Moving VMs and virtual appliances between clouds

Benefits: Migration, Hybrid Clouds, Disaster Recovery, Cloudbursting
Deployment Mode Considerations: When moving a VM out of one cloud and into another as two separate actions, conceivably two different ID management systems could be used. When moving VMs in a truly hybrid cloud, however, federated ID management standards will be needed.
Standardizations Needed: Common VM description format
Possible Standards: OVF from DMTF; OpenID, Oauth

6.4.3 Use Case: Selecting the best IaaS cloud vendor, public or private

Benefits: Provide cost-effective reliable deployments
Deployment Mode Considerations: When considering hybrid or distributed (inter)cloud deployments, uniform and consistent resource, performance, and policy descriptions are needed.
Standardizations Needed: Resource and performance requirements description languages.
Possible Standards: For basic resource descriptions, DMTF CIM and OGF GLUE are candidates. Other, more extensive description languages for performance or policy enforcement are to be determined.

6.4.4 Use Case: Portable tools for monitoring and managing clouds

Benefits: Simplifies operations as opposed to individual tools for each cloud
Deployment Mode Considerations: Monitoring and managing are separate but closely related tasks. The standards required will differ depending on whether the monitoring and managing must be done across trust boundaries or across distributed environments.
Standardizations Needed: Standard monitoring and management interfaces to IaaS resources
Possible Standards: Basic monitoring standards exist, such as the Syslog Protocol (IETF RFC 5424), which can be used with the Transport Layer Security (TLS) Transport Mapping for Syslog (IETF RFC 5425). Basic management standards include the Cloud Management WG from DMTF, and OCCI from OGF.

6.4.5 Use Case: Moving data between clouds

Benefits: Migration between Clouds, cross-cloud application and B2B integration
Deployment Mode Considerations: Migrating data from one cloud to another in two separate moves through the client is a simpler case. Migrating data directly from one cloud to another will require standards for federated identity, delegation of trust, and secure, third-party data transfers.
Standardizations Needed: Standard metadata/data formats for movement between clouds
Standardized query languages (e.g., for NoSQL for IaaS)
Possible Standards: AS4, OAGIS, NoSQL, GridFTP

6.4.6 Use Case: Single sign-on access to multiple clouds

Benefits: Simplified access, Cross-cloud applications
Deployment Mode Considerations: Single sign-on can mean using the same credentials to access different clouds independently at different times. Single sign-on to access an inter-cloud application that spans multiple clouds will require federated identity management, delegation of trust, and virtual organizations.
Standardizations Needed: Federated identity, authorization, and virtual organizations
Possible Standards: OpenID, OAuth, SAML, WS-Federation and WS-Trust, CSA outputs; Virtual Organization Management System (VOMS) is under development at OGF.

6.4.7 Use Case: Orchestrated processes across clouds and Enterprise Systems

Benefits: Direct support for necessarily distributed systems
Deployment Mode Considerations: This use case is inherently distributed and across trust boundaries. This can be generally termed *federated resource management* and is a central concept in the grid computing community. The term *inter-cloud* can also be used to denote this concept.
Standardizations Needed: To address this use case completely, an entire set of capabilities need to be standardized, e.g.,
- Infrastructure services ;
- Execution Management services ;
- Data services ;
- Resource Management services ;
- Security services;
- Self-management services; and
- Information services.

Possible Standards: SOA standards (such as WS-I) and grid standards (such as the OGSA WSRF Basic Profile, OGF GFD-R-P.072) exist that cover these areas, but issues around stateful resources, callbacks/notifications, and remote content lifetime management has caused these to be eclipsed by the simplicity of Representational State Transfer (REST). Hence, standard, REST-based versions of these capabilities

must be developed. Such work is being done in several organizations, including the IEEE.
- *DMTF and OGF*. The OGF Distributed Computing Infrastructure Federations Working Group (DCI Federal [DCIfed]-WG) is addressing two usage scenarios: (1) delegation of workload from one domain into the other, covering job description, submission, and monitoring; and (2) leasing of resources, including resource definition, provisioning, and monitoring. Existing standards to support this include WS-Agreement, Job Submission Description Language, GLUE, OGSA Basic Execution Service, OCCI, and Usage Record. Specific business application data formats may be supported by OAGIS.
- *Workflow and workflow engines* will also need standardization and adoption in the cloud arena. BPEL is one existing standard but extensions might be needed to efficiently support scientific and engineering workflows.

6.4.8 Use Case: Discovering cloud resources

Benefits: Selection of appropriate clouds for applications

Deployment Mode Considerations: To support inter-cloud resource discovery, secure federated catalog standards are needed.

Standardizations Needed: Description languages for available resources, Catalog interfaces

Possible Standards: This use case actually requires two areas of standardization: (1) description languages for the resources to be discovered, and (2) the discovery APIs for the discovery process itself. Some existing standards and tools cover both areas. RDF is a standard formalism for describing resources as triples consisting of subject-predicate-object. The Dublin Core is a small, fundamental set of text elements for describing resources of all types. It is commonly expressed in RDF. Since the Dublin Core is a "core" set, it is intended to be extensible for a broad range of application domains.

Such work is being pursued by the Dublin Core Metadata Initiative. ebXML Registry Information Model (ebRIM) actually defines both a description language and a discovery method, ebXML Registry Services (ebRS).

ID-WSF also defines both a discovery information model and discovery services that cover federated identity and access management. LDAP is an existing standard that has been used to build catalogue and discovery services, but issues might occur with regards to read vs. write optimization. UDDI is another existing standard from OASIS. A third existing standard is CSW from OGC that uses ebRIM. While this was originally developed to support geospatial applications, it is widely used in distributed catalogues that include services. All of these existing standards need to be evaluated for suitability for cataloguing and discovery of cloud resources and services.

6.4.9 Use Case: Evaluating SLAs and penalties

Benefits: Selection of appropriate cloud resources
Deployment Mode Considerations: SLAs will be primarily established between a single client and provider, and should observe any required standards for authentication, authorization, and non-repudiation. The need for SLAs between a single client but across multiple providers will be much less common. The difficulty in effectively implementing distributed SLAs will also discourage their development.
Standardizations Needed: SLA description language
Possible Standards: WS-Agreement (GFD.107) defines a language and a protocol for advertising the capabilities of service providers and creating agreements based on creational offers, and for monitoring agreement compliance at runtime. This is supported by WS-AgreementNegotiation (OGF), which defines a protocol for automated negotiation of offers, counter-offers, and terms of agreements defined under WS-Agreement-based service agreements.

6.4.10 Use Case: Auditing clouds

Benefits: Ensure regulatory compliance. Verify information assurance.
Deployment Mode Considerations: Auditing will be done primarily between a single client and provider, and should observe any required standards for authentication, authorization, integrity, and non-repudiation.
Standardizations Needed: Auditing standards and verification check lists
Possible Standards: CSA Cloud Audit. Relevant informational work can be found in *Guidelines for Auditing Grid Certificate Authorities* (OGF GFD.169).

Ongoing Roadmap analysis should track the development of the standards and update the Standards Inventory as necessary.

7 Cloud Computing Standards Gaps and USG Priorities

Cloud computing is the result of evolutions of distributed computing technologies, enabled by advances in fast and low-cost network, commoditized faster hardware, practical high-performance virtualization technologies, and maturing interactive Web technologies. Cloud computing continues to leverage the maturity of these underlying technologies, including a lot of standard-based technologies and system architecture components. As the previous clauses of the cloud computing standards survey show, the majority of cloud-relevant standards are from these pre-cloud era technologies.

In the meantime, there are emerging challenges in some areas in cloud computing that have been addressed by technology vendors and service providers' unique innovations. New service model interactions and the distributed nature in resource control and ownership in cloud computing have resulted in new standards gaps. Additionally, some pre-cloud computing era technology standardization gaps are being brought to the forefront by cloud computing. In summary, some areas of these gaps are introduced by new service model interactions and the distributed nature in resource control and ownership in cloud computing; some are pre-cloud computing era technology standardization gaps that are brought to the forefront.

In this clause, firstly, we use the cloud computing conceptual model from NIST Cloud Computing Reference Architecture and Taxonomy Working Group as described in Chapter 3 as the framework of reference to identify these gaps in need of standardization. Secondly, we use a broad set of USG business use cases as described in previous clauses and from NIST Cloud Computing Target Business Use Case Working Group, to identify priorities of standardization that will maximize the benefits and meet the more urgent needs of government consumers.

The following table summarizes the areas of standardization gaps and standardization priorities based on USG cloud computing adoption requirements.

Area of Standardization Gaps	Priorities for Standardization Based On USG Requirements
SaaS Functional Interfaces, e.g., - Data format and interface standards for email and office productivity - Metadata format and interface standards for e-discovery	High priorities on: - SaaS application specific data and metadata format standards to support interoperability and portability requirement when migrating high-value, low-risk applications to SaaS.
SaaS Self-Service Management Interfaces, e.g., - Interface standards related to user account and credential management	

Area of Standardization Gaps	Priorities for Standardization Based On USG Requirements
PaaS Functional Interfaces, e.g., - Standards of data format to support database serialization and de-serialization	
Business Support, Provisioning and Configuration, e.g., - Standards for describing cloud service-level agreement and quality of services - Standards for describing and discovering cloud service resources - Standards for metering and billing of service consumptions and usage	High priorities on: - Resource description and discovery standards to support data center consolidation using private and community IaaS clouds
Security and Privacy, e.g., - Standards for identity provisioning and management across different network and administration domains - Standards for secure and efficient replication of identity and access policy information across systems - Single Sign-On (SSO) interface and protocol standards that support strong authentication - Standards in policies, processes, and technical controls in supporting the security auditing, regulation, and law compliance needs	High priorities on: - Security auditing and compliance standards to support secure deployment, assess, and accreditation process for cloud-specific deployment - Identity and access management standards to support secure integration of cloud systems into existing enterprise security infrastructure

Table 8 – Areas of Standardization Gaps and Standardization Priorities

7.1 Areas of Standardization Gaps

As the cloud computing conceptual model indicates, cloud computing consumers do not have direct visibility into the physical computing resources. Instead, consumers interact with service providers through three service model interfaces, namely, IaaS, PaaS, and SaaS, to gain a view to the abstracted computing resource they are renting. As described in Chapter 5, these interaction interfaces can be categorized into two types: (1) functional interfaces that expose the primary function of the service, and (2) management interfaces that let the consumers to manage the rented computing resources. The following areas of standardization gaps are observed through the standards inventory:

7.1.1 SaaS Functional Interfaces

The varieties of the SaaS applications determine *what* can be consumed by the SaaS consumer. There are varying degrees of functional standardization. SaaS applications are mostly consumed using a Web browser, and some are consumed as a Web service using other application clients, such as standalone desktop applications and mobile applications. Even as most SaaS applications are using Web and Web service standards to deliver these application capabilities, application-specific data and metadata standards remain an area of standardization gaps in portability and interoperability. For example, email and office productivity application data format standards and interfaces are required to achieve interoperability and portability for migrating from existing systems to cloud-based systems. Another important area for standardization is the metadata format and interfaces, in particular, to support compliance needs. For example, standard metadata format and APIs to describe and generate e-discovery metadata for emails, document management systems, financial account systems, etc., that will help government consumers to leverage commercial off-the-shelf (COTS) and government off-the-shelf (GOTS) software products to meet e-discovery requirements. This is especially important when email messaging systems, content management systems, or Enterprise Resource Planning (ERP) financial systems are migrated to a SaaS model.

7.1.2 SaaS Self-service Management Interfaces

Due to the diverse domain and functional differences among SaaS offerings, the management interfaces used for the consumers to administer and customize the application functionalities are also very diverse. However, certain management functionalities are common, such as those related to user account and credential management. These common management functionalities represent candidates for interoperability standardization.

7.1.3 PaaS Functional Interfaces

PaaS functional interfaces encompass the runtime environment with supporting libraries and system components for developers to develop and deploy SaaS applications. Standard-based APIs are often part of a PaaS offering to begin with (such that the PaaS provider can lure existing development away to cloud-based hosting environment). However, data format for backup and migration of application workload, including database serialization/de-serialization, need further standardization to support portability.

7.1.4 Business Support, Provisioning and Configuration

In cloud service management areas, the importance of standard data formats and interfaces to describe service-level agreement (SLA) and quality of service (QoS) in traditional IT systems is high. While standards do exist for SLA negotiation and automated service condition matching, the application of these to the fine level of detail expected for large-scale cloud use cases is just developing. Computing resource description and discovery are also a forefront area in need of standardization as consumers transition from buying and managing resources to renting resources in a cloud environment. This is limited not only to raw computing resources such as virtualized processing, storage, and networking resources, but also includes higher-level abstractions of application processing resources. A standardization gap identified in a related area is metering and billing of service consumptions; data formats and management interfaces are used to report, deliver and communicate this usage information.

7.1.5 Security and Privacy

As cloud systems are typically external components in the consumer organizations overall IT system, especially in the out-sourced (off-site) deployment models, the need to have seamless security integration calls for interoperable standard interfaces for authentication, authorization, and communication protections. The challenges of identity and access management across different network and administration domains are more prominent in the cloud environment as the implementation of these capabilities within the cloud systems is often not the same organization as consumer organization where the identity information originates. Standardization in areas such as identity provisioning, management, secure and efficient replication across different systems, and identity federation will greatly help to improve the identity management capabilities in the cloud. A related area with specifically wide government usage that can benefit from standardization is single sign-on interface and protocols that support strong authentication.

Government IT systems require strong auditing and compliance needs. In a lot of cases, these requirements must be in place before a system can be approved for operation. This is another area that requires standardization and is exacerbated as the consumer organizations typically do not own or control the underlying system resources that implement the system capabilities. Standardization in policies, processes, and technical controls in supporting the security auditing requirements, regulations, and law compliance needs to consider the collaboration process between the cloud consumers and providers, their roles and the sharing of the responsibilities in implementing these capabilities.

7.2 Standardization Priorities Based on USG Priorities to Standardization Priorities Based on USG Cloud Computing Adoption Priorities

As described in the Federal Cloud Computing Strategy, there are cloud computing business use cases that have higher priorities than others. The requirements expressed in these high-priority target business use cases can be used to prioritize the standardization gaps. For example, various USG groups have identified data center consolidation using virtualization technologies as one of the primary goals in the next few years. Migrating collaboration applications, including email messaging (email, contacts, and calendars) and online office productivity application, to the cloud is also quoted as an early target of government cloud operation.

By analyzing the USG cloud computing target business use cases with their specific technical requirements, one can point out the following basic drivers that can be used to prioritize cloud computing standard gaps:

- The focus on supporting migration of system workload, including data, metadata and processing logic of existing in-house IT systems, to cloud-based systems to ensure continuous operation; this focus is centered on portability standards.

- The need to have interoperability between existing in-house IT systems and cloud-based systems, as cloud-deployed systems will not be the only part of the overall enterprise system; this need is centered on interoperability standards, including security interoperability standards.

- The need to help government consumers to choose and buy the most cost-effective solutions. If a cloud solution is not as economical as an in-house traditional IT system, there is no financial incentive to move the system to the cloud.

Based on these understandings, the following areas of standardization gaps in cloud computing are of higher priorities for USG cloud consumers:

7.2.1 Security Auditing and Compliance

Auditing and compliance data and metadata format standards are needed. Standard interfaces to retrieve and manage these data and metadata assets are also required to be integrated with existing tools and processes. In addition, policy, process and technical control standards are needed to support more manageable assessment and accreditation processes, which are often a prerequisite before a system is put in operation.

7.2.2 Identity and Access Management

As described earlier, security integration of a cloud-based system into existing enterprise security infrastructure is a must for the majority of government systems with moderate and greater impact. Existing practices of external cloud-based components in identity and access management is often based on proprietary and custom integration solutions. Constant and standard ways of provisioning identity data, managing identity data, and replicating to-and-from cloud-based system components, are needed to ensure that consumer organizations' short-term and long-terms needs are met.

A lot of government systems are required to have strong authentication, such as two-factor authentication implemented in an Internet-deployed system. Standards in supporting single sign-on and strong authentication are a must for these types of systems.

7.2.3 SaaS Application Specific Data and Metadata

To support the urgent need to migrate certain applications to the cloud, application-specific data and metadata format standards are required. This is an area where a lot of SaaS providers currently help consumer organizations to migrate their existing system by offering custom conversion and migration support. However, without standards in data and metadata format for these applications, there is the potential danger of creating non-interoperable islands of cloud solutions and vendor lock-in. For example, some SaaS email solutions may not be fully interoperable with in-house email and calendaring solutions. There are specific email working groups in the federal cloud computing initiative that are looking into putting forward specific metadata standardization requirements for email security, privacy, and record management. Other SaaS functional areas, such as document management and financial systems are also among the high-priority areas where standards in data and metadata are needed.

7.2.4 Resource Description and Discovery

Descriptions and discovery of computing resources needs are usually the first steps for consumers to take to start using cloud computing. Standard ways of resource descriptions will facilitate programmatically developing interoperable cloud applications to discover and use cloud computing resources, be it computing resources, storage resources, or application resources. In establishing private or community cloud computing as a way to implement data center consolidation, standards for these areas are important to not only help avoid implementing vendor-specific interfaces, but also to help increase the dynamic provisioning capabilities of the solution and utility of the computing resources.

8 Conclusions and Recommendations

8.1 Conclusions

Cloud computing can enable USG agencies to achieve cost savings and increased ability to quickly create and deploy enterprise applications. While cloud computing technology challenges many traditional approaches to datacenter and enterprise application design and management, requirements for interoperability, portability, and security remain critically important for successful deployments. Technically sound and timely standards are key to ensuring that requirements for interoperability, portability, and security are met.

There is a fast-changing landscape of cloud computing-relevant standardization under way in a number of SDOs. While there are only a few approved cloud computing-specific standards at present, USG agencies should be encouraged to participate in specific cloud computing standards development projects that support their priorities in cloud computing services.

8.2 Recommendations for Accelerating the Development and Use of Cloud Computing Standards

Recommendation 1 – Contribute Agency Requirements
Agencies should contribute clear and comprehensive user requirements for cloud computing standards projects.

Recommendation 2 – Participate in Standards Development
Agencies should participate in the cloud computing standards development process at the highest possible level that is commensurate with their need and available resources for a particular standard. The level of participation can be broadly categorized as follows:

Monitor: Provide the necessary resources to monitor and report on SDO activities on the standard of interest. Monitoring will require some level of technical expertise and communication for internal stakeholders.

Influence: Provide the necessary resources to attend meetings, present use cases, and guide the developing standard. This will require a uniform set of agency requirements to be presented to all relevant SDOs.

Promote: Provide the necessary resources for agencies to engage, either individually or jointly with other agencies, in prototyping projects to promote the development and adoption of critical standards, the demonstration of interoperable implementations, and the demonstration of integrated, end-to-end capabilities based on standard tooling.

Lead: Provide the necessary resources for agency personnel to take leadership positions within SDOs to lead the development and adoption of needed standards. This will require technical expertise and leadership skills.

Recommendation 3 – Encourage Compliance Testing to Accelerate Technically Sound Standards-Based Deployments
Agencies should support the concurrent development of conformity and interoperability assessment schemes to accelerate the development and use of technically sound cloud computing standards and standards-based products, processes, and services.

Recommendation 4 – Specify Cloud Computing Standards
Agencies should specify cloud computing standards in their procurements and grant guidance when multiple vendors offer standards-based implementations and there is evidence of successful interoperability testing. In such cases, agencies should ask vendors to show compliance to the specified standards.

Recommendation 5 –USG – Wide Use of Cloud Computing Standards
To support USG requirements for interoperability, portability, and security in cloud computing, the Federal Standards and Technology Working Group chaired by NIST and complimentary to the Fed CIO Council Cloud Computing Executive Steering Committee (CCESC) and Cloud First Task Force should recommend specific cloud computing standards and best practices for USG-wide use.

Recommendation 6 – Dissemination of Information on Cloud Computing Standards
A listing of standards relevant to cloud computing should be posted and maintained by NIST.

Bibliography

This section provides sources for additional information.

Distributed Management Task Force (DMTF)

- Interoperable Clouds White Paper

DSP-IS0101 Cloud Interoperability White Paper V1.0.0
This white paper describes a snapshot of the work being done in the DMTF Open Cloud Standards Incubator, including use cases and reference architecture as they relate to the interfaces between a cloud service provider and a cloud service consumer.
http://dmtf.org/sites/default/files/standards/documents/DSP-IS0101_1.0.0.pdf

- Architecture for Managing Clouds White Paper

DSP-IS0102 Architecture for Managing Clouds White Paper V1.0.0
This white paper is one of two Phase 2 deliverables from the DMTF Cloud Incubator and describes the reference architecture as it relates to the interfaces between a cloud service provider and a cloud service consumer. The goal of the Incubator is to define a set of architectural semantics that unify the interoperable management of enterprise and cloud computing. http://dmtf.org/sites/default/files/standards/documents/DSP-IS0102_1.0.0.pdf

- Use Cases and Interactions for Managing Clouds White Paper

DSP-IS0103 Use Cases and Interactions for Managing Clouds White Paper V1.0.0
This document is one of two documents that together describe how standardized interfaces and data formats can be used to manage clouds. This document focuses on use cases, interactions, and data formats. http://dmtf.org/sites/default/files/standards/documents/DSP-IS0103_1.0.0.pdf

Global Inter-Cloud Technology Forum (GICTF)
Use Cases and Functional Requirements for Inter-Cloud Computing
Published on August 2010
http://www.gictf.jp/doc/GICTF_Whitepaper_20100809.pdf

This whitepaper describes three areas of advantages of inter-cloud computing, which are assured or prioritized performance, availability, and convenience of combined services. Several use cases of inter-cloud computing are provided with details according to these three areas, such as assured performance against transient overload, disaster recovery and service continuity for availability, and federated service provisions, followed by sequential procedures, and functional requirements for each use case. Essential functional entities and interfaces are identified to meet these described requirements.

Appendix A – NIST Special Publications relevant to Cloud Computing

NIST Special Publication 800-125, *Guide to Security for Full Virtualization Technologies*

NIST Special Publication 800-144, *DRAFT Guidelines on Security and Privacy Issues in Public Cloud Computing*

NIST Special Publication 800-145, *DRAFT A NIST Definition of Cloud Computing*

NIST Special Publication 800-146, *DRAFT Cloud Computing Synopsis and Recommendations*

Appendix B – Definitions

Data Migration – The periodic transfer of data from one hardware or software configuration to another or from one generation of computer technology to a subsequent generation. Migration is a necessary action for retaining the integrity of the data and for allowing users to search, retrieve, and make use of data in the face of constantly changing technology.
[SOURCE : http://www.ischool.utexas.edu/~scisco/lis389c.5/email/gloss.html]

Information Technologies (IT) – Encompasses all technologies for the capture, storage, retrieval, processing, display, representation, organization, management, security, transfer, and interchange of data and information.
[SOURCE: This report]

Interoperability – The capability to communicate, execute programs, or transfer data among various functional units under specified conditions. [SOURCE: American National Standard Dictionary of Information Technology (ANSDIT)]

Maintainability – A measure of the ease with which maintenance of a functional unit can be performed using prescribed procedures and resources. Synonymous with serviceability.
[SOURCE: American National Standard Dictionary of Information Technology (ANSDIT)]

Network Resilience – A computing infrastructure that provides continuous business operation (i.e., highly resistant to disruption and able to operate in a degraded mode if damaged), rapid recovery if failure does occur, and the ability to scale to meet rapid or unpredictable demands.
[SOURCE: The Committee on National Security Systems Instruction No 4009,"National Information Assurance Glossary." CNSSI-4009]

Portability – The capability of a program to be executed on various types of data processing systems with little or no modification and without converting the program to a different language. [SOURCE: American National Standard Dictionary of Information Technology (ANSDIT)]

Portability – 1) The ability to transfer data from one system to another without being required to recreate or reenter data descriptions or to modify significantly the application being transported. 2) The ability of software or of a system to run on more than one type or size of computer under more than one operating system.
[SOURCE: Federal Standard 1037C, Glossary of Telecommunication Terms, 1996]

Privacy – Information privacy is the assured, proper, and consistent collection, processing, communication, use, and disposition of personal information (PI) and personally identifiable information (PII) throughout its life cycle.
[SOURCE: NIST Cloud Computing Reference Architecture and Taxonomy Working Group]

Reference implementation – An implementation of a standard to be used as a definitive interpretation for the requirements in that standard. Reference implementations can serve many purposes. They can be used to verify that the standard is implementable, validate conformance test tools, and support interoperability testing among other implementations. A reference implementation may or may not have the quality of a commercial product or service that implements the standard.
[SOURCE: This report]

Reliability – A measure of the ability of a functional unit to perform a required function under given conditions for a given time interval.
[SOURCE: American National Standard Dictionary of Information Technology (ANSDIT)]

Resilience - The ability to reduce the magnitude and/or duration of disruptive events to critical infrastructure. The effectiveness of a resilient infrastructure or enterprise depends upon its ability to anticipate, absorb, adapt to, and/or rapidly recover from a potentially disruptive event.
[SOURCE: CRITICAL INFRASTRUCTURE RESILIENCE FINAL REPORT AND RECOMMENDATIONS, NATIONAL INFRASTRUCTURE ADVISORY COUNCIL, SEPTEMBER 8, 2009]

Resilience – The adaptive capability of an organization in a complex and changing environment.
[SOURCE: ASIS International, ASIS SPC.1-2009, American National Standard, Organizational Resilience: Security, Preparedness, and Continuity Management System – Requirements with Guidance for Use.]

Security – Refers to information security. Information security means protecting information and information systems from unauthorized access, use, disclosure, disruption, modification, or destruction in order to provide:

- **Integrity,** which means guarding against improper information modification or destruction, and includes ensuring information non-repudiation and authenticity;
- **Confidentiality,** which means preserving authorized restrictions on access and disclosure, including means for protecting personal privacy and proprietary information; and
- **Availability,** which means ensuring timely and reliable access to and use of information.

[SOURCE: Title III of the E-Government Act, entitled the Federal Information Security Management Act of 2002 (FISMA)]

Standard – A document, established by consensus and approved by a recognized body that provides for common and repeated use, rules, guidelines or characteristics for activities or their results, aimed at the achievement of the optimum degree of order in a given context. Note: Standards should be based on the consolidated results of science, technology, and experience, and aimed at the promotion of optimum community benefits. [SOURCE: ISO/IEC Guide 2:2004, Standardization and related activities – General Vocabulary, definition 3.2]

Standard – A document **that** may provide the requirements for: a product, process or service; a management or engineering process; or a testing methodology. An example of a product standard is the multipart ISO/IEC 24727, *Integrated circuit card programming interfaces*. An example of a management process standard is the ISO/IEC 27000, *Information security management systems*, family of standards. An example of an engineering process standard is ISO/IEC 15288, System life cycle processes. An example of a testing methodology standard is the multipart ISO/IEC 19795, *Biometric Performance Testing and Reporting*.

Standards Developing Organization (SDO) – Any organization that develops and approves standards using various methods to establish consensus among its participants. Such organizations may be: accredited, such as ANSI-accredited IEEE; or international treaty-based, such as the ITU-T; or international private sector-based, such as ISO/IEC; or an international consortium, such as OASIS or IETF; or a government agency. SOURCE: [This report]

Usability – The extent to which a product can be used by specified users to achieve specified goals with effectiveness, efficiency, and satisfaction in a specified context of use. [SOURCE: ISO 9241-11:1998 Ergonomic requirements for office work with visual display terminals (VDTs) – Part 11: Guidance on usability and ISO/IEC 25062:2006 Software engineering – Software product Quality Requirements and Evaluation (SquaRE) – Common Industry Format (CIF) for usability test reports]

Appendix C – Acronyms

ANSDIT	American National Standard Dictionary of Information Technology
API	Application Programming Interface
BOD	Business Object Document
CCESC	Cloud Computing Executive Steering Committee
CDMI	Cloud Data Management Interface
CDN	Content Delivery Network
CIO	Chief Information Officer
CMWG	Cloud Management Working Group
COTS	Commercial Off-the-shelf
CPU	Central Processing Unit
CRM	Customer Relationship Management
CRUD	Create-Read-Update-Delete
CSA	Cloud Security Alliance
CSIRT	Computer Security Incident Response Teams
CSW	Catalog Service for the Web
DCIFed	DCI Federation Working Group
DISR	Defense IT Standards Registry
DMTF	Distributed Management Task Force
DoD	Department of Defense (USA)
ebRIM	Electronic business Registry Information Model
ebXML	Electronic Business using eXtensible Markup Language
ERP	Enterprise Resource Planning
EULA	End User License Agreement
FCCI	Federal Cloud Computing Initiative
FEA	Federal Enterprise Architecture
FIPS	Federal Information Processing Standards
GEIA	The Government Electronics & Information Technology Association
GICTF	Global Inter-Cloud Technology Forum
GLUE	Grid Laboratory Uniform Environment
GOTS	Government off-the-shelf
HTML	HyperText Markup Language
HTTP	Hypertext Transfer Protocol
ID-WSF	IDentity Web Service Framework
I/O	Input/Output
IaaS	Cloud Infrastructure as a Service
IEC	International Electrotechnical Commission

IEEE	Institute of Electrical and Electronic Engineers
IETF	Internet Engineering Task Force
IODEF	Incident Object Description Format
IP	Internet Protocol
ISIMC	Information Security and Identity Management Committee
ISO	International Organization for Standardization
ISO/IEC JTC 1	International Organization for Standardization/International Electrotechnical Commission Joint Technical Committee 1 Information Technology
IT (ICT)	Information Technology (Note: it is often referred to as ICT [Information and Communications Technologies])
ITU	International Telecommunication Union
ITU-T	The ITU Telecommunication Standardization Sector
J2EE	Java 2 Platform, Enterprise Edition
JSON	JavaScript Object Notation
KMIP	Key Management Interoperability Protocol
LDAP	Lightweight Directory Access Protocol
MID	mobile Internet devices (USA)
MIL-STDS	Military Standards (USA)
NIEM	National Information Exchange Model
NIST	National Institute of Standards and Technology
NIST SP	NIST Special Publication
OAGi	Open Applications Group
OAGIS	Open Applications Group Integration Specification
OASIS	Organization for the Advancement of Structured Information Standards
OAuth	Open Authorization Protocol
OCC	Open Cloud Consortium
OCCI	Open Cloud Computing Interface
ODF	Open Document Format
OGC	Open Geospatial Consortium
OGF	Open Grid Forum
OGSA	Open Grid Services Architecture
OMG	Object Management Group
OOXML	Office Open XML
OS	Operating System
OVF	Open Virtualization Format
P2P	Peer-to-Peer
PaaS	Cloud Platform as a Service
PDA	Personal Digital Assistant

PHP	PHP: Hypertext Preprocessor
PI	Personal Information
PII	Personal Identifiable Information
PIV	Personal Identity Verification
PKI	public key infrastructure
QoS	Quality of Service
RDF	Resource Description Framework
REST	Representational State Transfer
RSS	Really Simple Syndication
SaaS	Cloud Software as a Service
SAJACC	Standards Acceleration to Jumpstart Adoption of Cloud Computing
SAML	Security Assertion Markup Language
SCAP	Security Content Automation Protocol
SDOs	Standards Developing Organizations
SLA	Service Level Agreement
SNIA	Storage Networking Industry Association
SOA	service-oriented architecture
SOAP	Simple Object Access Protocol
SPML	Service Provisioning Markup Language
SSL	Secure Sockets Layer
SSO	Standard Setting Organization
STANAGS	Standardization Agreements
TCG	Trusted Computing Group
TCP	Transmission Control Protocol
TLS	Transport Layer Security
UDDI	Universal Description Discovery and Integration
USG	United States Government
VM	Virtual Machine
W3C	World Wide Web Consortium
WG	Working Group
XACML	OASIS eXtensible Access Control Markup Language
XML	Extensible Markup Language

Appendix D – Standards Developing Organizations

Global Information and Communications Technologies (IT) standards are developed in many venues. Such standards are created through collaborative efforts that have a global reach, are voluntary, and are widely adopted by the marketplace across national borders. These standards are developed not only by national member-based international standards bodies, but also by consortia groups and other organizations.

In July 2009, a Wiki site for cloud computing standards coordination was established: cloud-standards.org. The goal of the site is to document the activities of the various SDOs working on cloud computing standards.

The following is a list of SDOs that have standards projects and standards relevant to cloud computing.

CloudAudit
The goal of CloudAudit is to provide a common interface and namespace that allows cloud computing providers to automate the Audit, Assertion, Assessment, and Assurance (A6) of their infrastructure (IaaS), platform (PaaS), and application (SaaS) environments and allow authorized consumers of their services to do likewise via an open, extensible, and secure interface and methodology.

CloudAudit is a volunteer cross-industry effort from the best minds and talent in cloud, networking, security, audit, assurance, and architecture backgrounds.

The CloudAudit/A6 Working group was officially launched in January 2010 and has the participation of many of the largest cloud computing providers, integrators, and consultants.

Distributed Management Task Force (DMTF)
Open Virtualization Format (OVF)
DSP0243 Open Virtualization Format (OVF) V1.1.0
OVF has been designated as ANSI INCITS 469 2010
This specification describes an open, secure, portable, efficient, and extensible format for the packaging and distribution of software to be run in virtual machines.

Open Cloud Standards Incubator
DMTF's Open Cloud Standards Incubator focused on standardizing interactions between cloud environments by developing cloud management use cases, architectures, and interactions. This work was completed in July 2010. The work has now transitioned to the Cloud Management Working Group.

Cloud Management Working Group (CMWG)
The CMWG will develop a set of prescriptive specifications that deliver architectural semantics as well as implementation details to achieve interoperable management of clouds between service requestors/developers and providers. This WG will propose a resource model that, at a minimum, captures the key artifacts identified in the use cases and interactions for managing clouds document produced by the Open Cloud Incubator.

Using the recommendations developed by DMTF's Open Cloud Standards Incubator, the Cloud Management Workgroup (CMWG) is focused on standardizing interactions between cloud environments by developing specifications that deliver architectural semantics and implementation details to achieve interoperable cloud management between service providers and their consumers and developers.

Institute of Electrical and Electronic Engineers (IEEE)
The IEEE Standards Association (IEEE-SA), a globally recognized standards-setting body within the IEEE, develops consensus standards through an open process that engages industry and brings together a broad stakeholder community. IEEE standards set specifications and best practices based on current scientific and technological knowledge. The IEEE-SA has a portfolio of over 900 active standards and more than 500 standards under development. Examples of IEEE's cybersecurity standards are the wireless local area network (WLAN) computer communication security standards (e.g., IEEE 802.11 series).

The Internet Engineering Task Force (IETF)
The Internet Engineering Task Force (IETF) issues the standards and protocols used to protect the Internet and enable global electronic commerce. The IETF develops cyber security standards for the Internet. Current activities include Public Key Infrastructure Using X.509 (PKIX), Internet Protocol Security (IPsec), Transport Layer Security (TLS), Secure Electronic Mail (S/MIME V3), DNS Security Extensions (DNSSEC), and Keying and Authentication for Routing Protocols (karp). Another IETF standard is the Incident Object Description Format (IODEF), which provides a framework for sharing information commonly exchanged by Computer Security Incident Response Teams (CSIRTs) about computer security incidents. IODEF is an underpinning for the National Information Exchange Model (NIEM), which enables jurisdictions to effectively share critical information on cyber incident management, security configuration management, security vulnerability management, etc.

International Organization for Standardization/International Electrotechnical Commission Joint Technical Committee 1 Information Technology (ISO/IEC JTC 1)
http://www.iso.org/iso/jtc1_home.html
ISO/IEC JTC 1, Information Technology, develops international IT standards for global markets. ISO and IEC are private sector international standards-developing organizations. In 1987, ISO and IEC established a joint Technical Committee by combining existing IT standards groups within ISO and IEC under a new joint Technical Committee, JTC 1. JTC 1 members are National Standards Bodies of different countries. Presently, there are 66 members. Approximately 2100 technical experts from around the world work within JTC 1. There are presently 18 JTC 1 Subcommittees (SCs) in which most of JTC 1 standards projects are being developed.

JTC 1 SC 27 (IT Security Techniques) is the one JTC 1 SC that is completely focused on cyber security standardization. SC 27 approved the establishment of a six-month study period (SP) that ended in April 2011. The purpose of the SP was to investigate the security requirements for cloud computing and what would be a feasible program of standards work to meet these requirements. The study period involves SC27 WG 1 (Information Security Management), WG 4 – Security Control and Services, and WG 5 – Identity Management, Privacy Technology and Biometrics. It is likely that SC 27 will proceed with some form of cloud work by October2011). Many other JTC 1 SCs are directly involved in specific standards critical to cyber security, including SC 6 (public key infrastructure [PKI] certificates), SC 7 (software and systems engineering), SC 17 (identification cards and related devices), SC 22 (programming languages, software environments and system software interfaces), and SC 37 (biometrics). In October 2009, JTC 1 established a new SC 38 for standardization in the areas of Web services, Service-Oriented Architecture (SOA), and cloud computing. SC38 initiated a Cloud Computing Study that will end in September 2011. The primary purpose of the study is to analyze cloud computing standardization activities and to recommend new SC38 cloud computing standardization projects.

ITU Telecommunication Standardization Sector (ITU-T)
The ITU-T develops international standards for the IT infrastructure including voice, data, and video. ITU-T established a Focus Group on Cloud Computing (FG Cloud) - http://www.itu.int/en/ITU-T/focusgroups/cloud/Pages/default.aspx. The charter of the FG Cloud is to investigate standards needed to support services/applications of cloud computing that make use of telecommunication networks, specifically to:

- identify potential impacts on standards development and priorities for standards needed to promote and facilitate telecommunication/IT support for cloud computing;
- investigate the need for future study items for fixed and mobile networks in the scope of ITU-T;
- analyze which components would benefit most from interoperability and standardization;
- familiarize ITU-T and standardization communities with emerging attributes and challenges of telecommunication/IT support for cloud computing; and
- analyze the rate of change for cloud computing attributes, functions, and features for the purpose of assessing the appropriate timing of standardization of telecommunication/IT in support of cloud computing.

The Focus Group is collaborating with the worldwide cloud computing communities (e.g., research institutes, forums, academia) including other SDOs and consortia. The ITU-T Study Groups involved in standards relevant to cloud computing include: SG-13 (Next Generation Networks) and SG-17 (Network Security).

Kantara Initiative
Kantara Initiative was established on April 20, 2009, by leaders of several foundations and associations working on various aspects of digital identity, aka "the Venn of Identity." It is intended to be a robust and well-funded focal point for collaboration to address the issues we

each share across the identity community: Interoperability and Compliance Testing; Identity Assurance; Policy and Legal Issues; Privacy; Ownership and Liability; UX and Usability; Cross-Community Coordination and Collaboration; Education and Outreach; Market Research; Use Cases and Requirements; Harmonization; and Tool Development.

Organization for the Advancement of Structured Information Standards (OASIS)
Founded in 1993, OASIS is a not-for-profit consortium. OASIS develops open standards for the global information society. The consortium produces Web services standards along with standards for security, e-business, and standardization efforts in the public sector and for application-specific markets. OASIS has more than 5,000 participants representing over 600 organizations and individual members in 100 countries. OASIS has a number of projects related to cloud computing including: ID Cloud, SSTC, WSSX, E- gov, and iD Trust Community of Practice. OASIS security, access, and identity policy standards relevant to cloud computing include: SAML, XACML, SPML, WS-Security Policy, and WS-Trust.

The Open Cloud Consortium (OCC)
OCC is a member-driven organization that develops reference implementations, benchmarks, and standards for cloud computing. The OCC operates cloud testbeds, such as the Open Cloud Testbed and the OCC Virtual Network Testbed. The OCC also manages cloud computing infrastructure to support scientific research, such as the Open Science Data Cloud.

Open Grid Forum (OGF) Open Grid Forum (OGF) is a leading standards developing organization operating in the areas of grid, cloud, and related forms of advanced distributed computing. The OGF community pursues these topics through an open process for development, creation, and promotion of relevant specifications and use cases.

OGF engages partners and participants throughout the international arena to champion architectural blueprints related to cloud and grid computing and the associated specifications to enable the pervasive adoption of advanced distributed computing techniques for business and research worldwide.

Advanced computing built on OGF standards enables organizations to share computing and information resources across department and organizational boundaries in a secure, efficient manner. Organizations throughout the world use production distributed architectures built on these features to collaborate in areas as diverse as scientific research, drug discovery, financial risk analysis, and product design. The capacity and flexibility of distributed computing enables organizations to solve problems that until recently were not feasible to address due to interoperability, portability, security, cost and data-integration constraints.

Clouds, grids, and virtualized distributed architectures reduce costs through automation and improved IT resource utilization and improve organizational agility by enabling more efficient business processes. OGF's extensive experience has enabled distributed computing built on these architectures to become a more flexible, efficient, and utility-like global computing infrastructure.

Standardization is the key to realizing the full vision and benefits of distributed computing. The standards developed by OGF enable the diverse resources of today's modern computing environment to be discovered, accessed, allocated, monitored, and managed as interconnected flexible virtual systems, even when provided by different vendors and/or operated by different organizations.

Open Cloud Computing Interface (OCCI) Working Group
The purpose of this group is the creation of a practical solution to interface with cloud infrastructures exposed as a service (IaaS). The group will focus on a solution which covers the provisioning, monitoring, and definition of cloud infrastructure services. The group should create this API in an agile way and overlapping work and efforts will be contributed and synchronized with other groups.

- Open Cloud Computing Interface Specification
- Open Cloud Computing Interface Terms and Diagrams

Object Management Group (OMG)
The OMG was founded in 1989 and develops standards for enterprise integration. Its membership is international and is open to any organization, both computer industry vendors and software end users. Specific cloud-related specification efforts have only just begun in OMG, focusing on modeling deployment of applications and services on clouds for portability, interoperability, and reuse.

Storage Networking Industry Association (SNIA)
SNIA Cloud TWG
The SNIA has created the Cloud Storage Technical Work Group for the purpose of developing SNIA Architecture related to system implementations of cloud storage technology. The cloud Storage TWG:

Acts as the primary technical entity for the SNIA to identify, develop, and coordinate systems standards for cloud storage.
Produces a comprehensive set of specifications and drives consistency of interface standards and messages across the various cloud storage-related efforts.
Documents system-level requirements and shares these with other cloud storage standards organizations under the guidance of the SNIA Technical Council and in cooperation with the SNIA Strategic Alliances Committee

SNIA Cloud Data Management Interface (CDMI)
The CDMI specification is now a SNIA Architecture standard and will be submitted to the INCITS organization for ratification as an ANSI and ISO standard as well.

SNIA Terms and Diagrams
SNIA and OGF have collaborated on cloud storage for a cloud computing whitepaper. A demo of this architecture has been implemented and shown several times. More information can be found at the Cloud Demo Google Group.

The Trusted Computing Group (TCG)
The TCG is a not-for-profit organization formed to develop, define, and promote open, vendor-neutral industry standards for trusted computing building blocks and software interfaces across multiple platforms. TCG has approximately 100 members from across the computing industry, including component vendors, software developers, systems vendors, and network and infrastructure companies.

World Wide Web Consortium (W3C)
Founded in 1994, the W3C is a non-incorporated international community of 334 Member organizations that develop standards in support of web technologies. The W3C work in the area of cyber security standards includes secure transferring data from one domain to another domain or between applications with well-defined document authentication. XML Encryption and XML Signature are key pieces of the XML security stack.

Appendix E – Conceptual Models and Architectures

General reference models:
- Distributed Management Task Force (DMTF): Cloud Service Reference Architecture
- Cloud Computing Use Case Discussion Group: a taxonomy for cloud computing
- IBM: Cloud Reference Architecture
- Cloud Security Alliance: Cloud Reference Model
- Cisco Cloud Reference Architecture Framework
- IETF: Cloud Reference Framework
- ITU-T Focus Group Cloud Reference Architecture

Reference models focusing on specific application requirements:
- Open Security Architecture: Secure Architecture Models
- GSA: FCCI (Federal Cloud Computing Initiative)
- Juniper Networks: Cloud-ready Data Center Reference Architecture
- SNIA standard: Cloud Data Management Interface
- Elastra: A Cloud Technology Reference Model for Enterprise Clouds

Appendix F – Examples of USG Criteria for Selection of Standards

USG Approach to Selecting Standards

F-1   USG Analysis Model for Selection of Private Sector Consensus Standards to be E-Gov Standards

The NIST E-Gov Standards Resource Center at Standards.gov includes the following list of questions that USG agencies can use when evaluating private sector consensus standards for agency use:

Applicability of standard

- Is it clear who should use the standard and for what applications?
- How does the standard fit into the Federal Enterprise Architecture (FEA)?
- What was done to investigate viable alternative standards (i.e., due diligence) before selecting this standard?

Availability of standard

- Is the standard published and publicly available?
- Is a copy of the standard free or must it be purchased?
- Are there any licensing requirements for using the standard?

Completeness of standard

- To what degree does the candidate standard define and cover the key features necessary to support the specific E-Gov functional area or service?

Implementations on standard

- Does the standard have strong support in the commercial marketplace?
- What commercial products exist for this standard?
- Are there products from different vendors in the market to implement this standard?
- Are there any existing or planned mechanisms to assess conformity of implementations to the standard?

Interoperability of implementations

- How does this standard provide users the ability to access applications and services through web services?
- What are the existing or planned mechanisms to assess the interoperability of different vendor implementations?

Legal considerations

- Are there any patent assertions made to this standard?
- Are there any IPR assertions that will hinder USG distribution of the standard?

Maturity of standard

- How technically mature is the standard?
- Is the underlying technology of the standard well-understood (e.g., a reference model is well-defined, appropriate concepts of the technology are in widespread use, the technology may have been in use for many years, a formal mathematical model is defined, etc.)?
- Is the standard based upon technology that has not been well-defined and may be relatively new?

Source of standard

- What standards body developed and now maintains this standard?
- Is this standard a de jure or de facto national or international standard?
- Is there an open process for revising or amending this standard?

Stability of standard

- How long has this standard been used?
- Is the standard stable (e.g., its technical content is mature)?
- Are major revisions or amendments in progress that will affect backward compatibility with the approved standard?
- When is the estimated completion date for the next version?

Department of Defense (DoD)

The DoD IT Standards Registry (DISR) mandates the minimum set of IT standards and guidelines for the acquisition of all DoD systems that produce, use, or exchange information. The Defense Information Systems Agency (DISA) is the executive agent for the DISR. The DISR is updated three times a year.

Initial Standards Selection Criteria for Inclusion in the DISR

A number of criteria should be considered when evaluating a standard for inclusion in the DISR. Selection criteria include:

- the source of the standard;
- openness;
- technology relevance;

- maturity;
- marketplace support;
- "usefulness/utility"; and
- risk.

Criteria	Description
Source of the Standard	Recognized authority
	Cooperative stance
	Feedback
	Process
	Consensus
Openness	Ownership/IPR
	User Participation
	Vendor Participation
Technology Relevance	
Maturity	Planning Horizon
	Stability
	Revision Content & Schedule
Marketplace Support	Acceptance
	Commercial Viability
Usefulness/Utility	Well-Defined Quality Attributes
	Services & Application Interoperability
Risk	Performance, maturity & stability issues

Table 9 – DoD Selection Criteria and Description Summary

Standards Source

DoD policy articulates a preference hierarchy based on the source (owner/sponsor/publisher) of the standard. Note that the 5th Priority, Military, has its own internal priority of international first and then DoD MIL-STDs.

The standards preference hierarchy is:

Priority	Standards Source Hierarchy	Example
1st	International	ISO, IEC, ITU
2nd	National	ANSI
3rd	Professional Society; Technology Consortia; Industry Association	IEEE; IETF; W3C; OASIS; GEIA
4th	Government	FIPS
5th	Military	MIL-STDS, STANAGS

Table 10 – DoD Standards Sources Preferences

The standard must be recognized as being available from a reputable and authoritative source. The responsible SDO/Standard Setting Organization (SSO) must have an established position within the relevant technical, professional, and marketplace communities as an objective authority in its sphere of activity. This means that the standard has been created and approved/adopted/published via a formal process and configuration management of the standard has been established. Accreditation implies acceptance by a recognized authoritative SSO.

The Standards Selection Criteria also provides guidance for moving through the standards life cycle that changes the category of a standard from "*emerging*" to "*mandated*" to "*inactive/retired*."

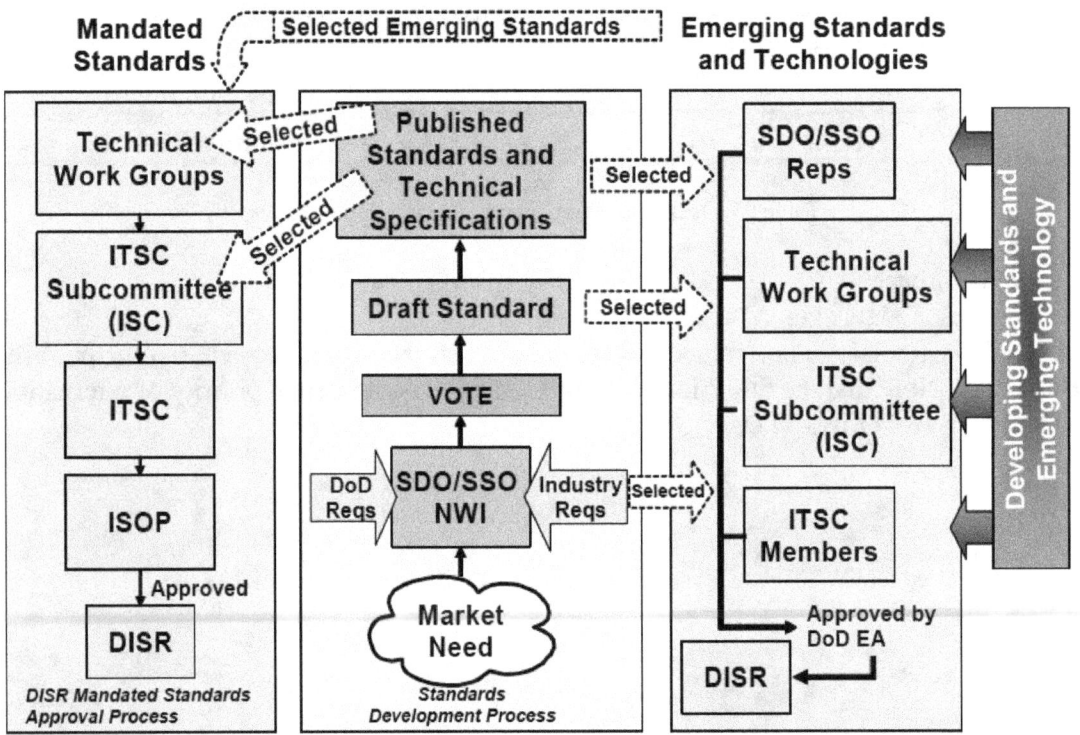

Figure 13 – DoD DISR Standards Selection Process

www.ingramcontent.com/pod-product-compliance
Lightning Source LLC
Chambersburg PA
CBHW081845170526
45167CB00007B/2911